Nuclear and Particle Physics

PHYSICS AND ITS APPLICATIONS

Series Editor

E.R. Dobbs
University of London

This series of short texts on advanced topics for students, scientists and engineers will appeal to readers seeking to broaden their knowledge of the physics underlying modern technology.

Each text provides a concise review of the fundamental physics and current developments in the area, with references to treatises and the primary literature to facilitate further study. Additionally texts providing a core course in physics are included to form a ready reference collection.

The rapid pace of technological change today is based on the most recent scientific advances. This series is, therefore, particularly suitable for those engaged in research and development, who frequently require a rapid summary of another topic in physics or a new application of physical principles in their work. Many of the texts will also be suitable for final year undergraduate and postgraduate courses.

Nuclear and Particle Physics

R.J. Blin-Stoyle FRS

CHAPMAN & HALL
London · New York · Tokyo · Melbourne · Madras

UK   Chapman & Hall, 2–6 Boundary Row, London SE1 8HN

USA   Van Nostrand Reinhold, 115 5th Avenue, New York NY 10003

JAPAN   Chapman & Hall Japan, Thomson Publishing Japan,
            Hirakawacho Nemoto Building, 7F, 1-7-11 Hirakawa-cho,
            Chiyoda-ku, Tokyo 102

AUSTRALIA   Chapman & Hall Australia, Thomas Nelson Australia,
            102 Dodds Street, South Melbourne, Victoria 3205

INDIA   Chapman & Hall India, R. Seshadri, 32 Second Main Road,
            CIT East, Madras 600 035

First edition 1991

© 1991 R.J. Blin-Stoyle

Typeset in 10/12 Times by
Thomson Press (India) Ltd, New Delhi

British Library Cataloguing in Publication Data

Blin-Stoyle, R.J. (Roger John)
  Nuclear and particle physics.
  1. Nuclear physics.
  I. Title II. Series
  539.7
  ISBN-13: 978-0-412-38320-5   e-ISBN-13: 978-94-010-9561-7
  DOI: 10.1007/978-94-010-9561-7

Library of Congress Cataloging-in-Publication Data

Blin-Stoyle, R.J. (Roger John)
  Nuclear and particle physics/R.J. Blin-Stoyle. – 1st ed.
    p cm. – (Physics and its applications; 4)
  Includes bibliographical references.

  1. Nuclear physics. 2. Particles (Nuclear physics) I. Title.
  II. Series.
  QC777.B58 1991   90-49711
  539.7 – dc20   CIP

Contents

Symbols and Notation

A	mass number
A	axial vector
$A(t)$	activity (radioactivity)
B	binding energy; also baryon number
b	barn
$\mathfrak{B}$	bottomness
Bq	Becquerel
C	charge conjugation operator
c	speed of light
$\mathfrak{C}$	charm
d	electric dipole moment
E	energy
e	unit of charge
eV	electronvolt
$F(Z, E)$	Fermi function
G	gravitational constant
G_A	axial vector β-decay coupling constant
G_F	Fermi constant $(= G_\mu)$
G_V	polar vector β-decay coupling constant
G_β	β-decay coupling constant
G_μ	μ-decay coupling constant
g	g-factor; also weak interaction coupling constant
g_s	strong interaction coupling constant
H	Hamiltonian operator
h	Planck's constant
$\hbar$	$h/2\pi$
I	isospin (see below) of nucleus or particle; also moment of inertia

I_3	3-component of isospin
J	spin (see below) of nucleus or elementary particle
$J(ij)$	weak current involving particles i and j
j	total angular momentum quantum number of a nucleon
L	orbital angular momentum operator
L_i	lepton number (i = e, μ, τ)
l	orbital angular momentum quantum number
M	atomic mass
m	z component of j
m_i	mass of particle i
m_l	z component of l
N	neutron number
N_A	Avogadro's number
n	radial quantum number
P	parity ($= \pm 1$); also probability
P	parity operator
p	momentum
Q	quadrupole moment; also Q value of a nuclear reaction; also electric charge of elementary particle
R	nuclear radius
S	strangeness; also separation energy
s_i	spin (see below) of nucleon (i = n, p) or quark (i = u, d, ...)
T	kinetic energy; also transition probability
t	time
$t_{1/2}$	half-life
$\mathfrak{T}$	topness
u	symbol for wave function
u	atomic mass unit
V	potential energy
V	polar vector
v	speed; also used as symbol for wave function
Y	hypercharge ($= B + S$)
$Y_{lm}(\theta\phi)$	spherical harmonic
Z	atomic number
α	fine structure constant (strength of electromagnetic interaction)
α_s	strength of strong interaction (cf. α)
Γ	width of resonance
ε	energy in units of $m_e c^2$
ε_0	permittivity of vacuum
θ_C	Cabbibo angle

θ_W	weak mixing angle
λ	wavelength (usually de Broglie wavelength); also decay constant; also helicity
$\lambdabar$	$\lambda/2\pi$
μ	magnetic moment; also reduced mass
μ_B	Bohr magneton
μ_N	nuclear magneton
ν	frequency
ρ	density (mass or charge)
σ	reaction cross-section
τ	mean life
ψ	wave function
Ω	solid angle
ω	angular frequency; also angular velocity

Note on spin and isospin. The notations used in nuclear physics and particle physics for spin and isospin tend to differ. Many books and publications use the following:

> nuclear physics
> > nuclear spin $\quad\quad I$
> > nuclear isospin $\quad T$
>
> particle physics
> > particle spin $\quad\quad J$
> > particle isospin $\quad I$

The confusion is obvious.

In this book, which covers both fields, the following notation will be used throughout:

> nuclear spin $\quad\quad J$ (operator $\mathbf{J}$)
> particle spin $\quad\quad J$ (operator $\mathbf{J}$)
> nuclear isospin $\quad I$ (operator $\mathbf{I}$)
> particle isospin $\quad I$ (operator $\mathbf{I}$)

One exception is that, when considering nuclear structure, the spins of the individual neutrons and protons will be denoted by s (operator $\mathbf{s}$). Similarly, when considering the quark structure of elementary particles the quark spins will be denoted by s (operator $\mathbf{s}$). In both cases appropriate suffices (e.g. n, p, u, d,...) will be added.

Preface

This book is intended to give a clear and concise introductory account of the basic ideas underlying nuclear and elementary particle physics. The attempt throughout is to convey a sound physical understanding of the structures and processes encountered. It assumes some knowledge of elementary quantum mechanics, particularly the treatment of angular momentum, and the rudiments of special relativity. In addition to 'standard' calculations based on this knowledge, frequent use is made of 'order-of-magnitude' and 'dimensional' arguments. In this way it has been possible to give some discussion of quite advanced topics and recent developments. Although reference is made from time to time to the apparatus of nuclear and particle physics no technical detail is given. My basic hope is that students using this book will acquire a sound understanding of what nuclear and particle physics is about and will wish to learn more.

I am indebted to Dr David Bailin and various (nameless) referees for penetrating and helpful comments on parts of the text.

Roger Blin-Stoyle

1

The nuclear atom

1.1 PROPERTIES OF ATOMS

The idea that matter consists of an assembly of atoms derived from Democritus (5th century BC). But it was not until the renaissance that development of the atomic concept began (e.g. by Hooke in the 17th century) and crude ideas about a kinetic description of matter were discussed. In the early 19th century Dalton and Gay-Lussac carried out quantitative chemical experiments which began to put the atomic hypothesis onto a firm basis as well as establishing the idea that atoms can combine together in the form of molecules. Following this, the kinetic theory of gases was developed in which physical properties of a gas such as pressure and temperature could be understood in terms of the motion of individual atoms and molecules. This culminated in the formulation of statistical thermodynamics in the middle of the 19th century by Maxwell, Boltzmann and Gibbs and the atomic concept became firmly established.

An atom, which is the smallest entity having the properties of an **element**, can be characterized by a number of features, in particular its chemical properties, its mass and its size.

1.1.1 Chemical Properties

Its chemical properties are reflected in the way in which it takes part in chemical reactions and the formation of molecules. These properties are unique to each atom, but there are groups of atoms having similar chemical properties and these regularities were embodied by Mendeleev (1872) in the periodic table of the elements. The groupings can be understood in terms of the detailed structure of the atom (section 1.3).

1.1.2 Atomic Mass

The work of Dalton (1808), Avogadro (1811) and subsequent more detailed studies of the way in which different atoms combine by mass and by volume to form molecules established that all atoms had a mass which was approximately an integer multiple of the mass of the hydrogen atom. It was also hypothesized by Avogadro (Avogadro's hypothesis or law) that at a given temperature and pressure equal volumes of a gas or vapour contain the same number of molecules. Unless the gas is monatomic (e.g. helium), the molecules will consist of two (diatomic) or more atoms bound together.

Following from the above it would be natural to specify the mass of hydrogen as 1 **atomic mass unit** and to specify the masses of other atoms in terms of this. Such indeed was the historical situation but now, in the SI system of units, the reference atom is taken to be carbon-12, the common carbon isotope in which the nucleus contains six protons and six neutrons (section 1.4), whose mass is taken to be exactly 12 atomic mass units (denoted by u), i.e.

$$\text{mass of one atom of carbon-12} = 12.000\ldots\text{u}$$

All atomic and molecular weights can then be expressed in terms of this unit and, on this basis, the mass of the hydrogen atom is 1.007 825 u.

A further definition is that of a **mole** which is equal to $M \times 10^{-3}$ kg of a substance whose molecular weight is M u. Since the mass of a mole is proportional to the molecular weight it follows that one mole of any substance contains the same number of molecules. This is known as Avogadro's number (N_A) and can be determined, for example, from electrolysis or X-ray diffraction experiments. It has the value

$$N_A = 6.0221 \times 10^{23}$$

Knowing N_A it is then possible to express 1 u in terms of kilograms. Thus, for carbon-12 we have **exactly** 0.012 kg contains N_A atoms of mass 12 u, i.e.

$$12 \times 6.0221 \times 10^{23}\,\text{u} = 0.012\,\text{kg}$$

or

$$1\,\text{u} = 1.6605 \times 10^{-27}\,\text{kg}$$

Using the Einstein relation between mass M and energy E $(E = Mc^2)$, the energy equivalent of 1 u is

$$E_u = 1\,\text{u} \times c^2 = 1.4924 \times 10^{-10}\,\text{J} = 931.49\,\text{MeV}$$

Note: 1 MeV = 10^6 eV where 1 eV (1 electronvolt) is the energy given to a particle of charge e (the magnitude of the electron charge = 1.602×10^{-19} C) when accelerated through an electrical potential difference of 1 V. Thus 1 MeV = 1.602×10^{-13} J.

Frequently the masses of atoms, nuclei and elementary particles are referred to in terms of their energy equivalents. In these terms 1 u is equivalent to 931.49 MeV/c^2.

Atomic masses vary from $\simeq 1$ u (for hydrogen) through to $\simeq 207$ u (for lead) and beyond (for the transuranic elements).

1.1.3 Atomic Size

Atomic sizes can be estimated approximately by assuming that in a solid the atoms are, roughly speaking, in contact. Taking an atom to have a spherical shape of radius R the separation between atoms will be $2R$ and each atom will occupy a volume of order $(2R)^3$. Consider now a substance whose atomic mass is M u. One mole will have mass $M \times 10^{-3}$ kg and, containing N_A atoms, will therefore occupy a total volume of $8N_A R^3$. If the density of the substance is ρ, this volume is also equal to $M \times 10^{-3}/\rho$ so that equating the two expressions gives

$$R = \frac{1}{2}\left(\frac{M \times 10^{-3}}{\rho N_A}\right)^{1/3}$$

In the case of gold, $\rho = 19.3$ kg m^{-3} and $M \simeq 197$, and from the above expression it follows that $R \simeq 1.3 \times 10^{-10}$ m. This is typical of atomic radii which, in the main, lie in the range $(0.5-2.5) \times 10^{-10}$ m.

1.2 RADIOACTIVITY

Towards the turn of the 19th century and into the 20th century it became clear that some heavy 'atoms' emitted various forms of radiation hitherto unknown. The idea of radiation in the form of beams of electrons had been demonstrated by J. J. Thomson (1895) through his studies using discharge tubes and the charge and mass of the electron were known ($e \simeq 1.6 \times 10^{-19}$ C, $m \simeq 9.1 \times 10^{-31}$ kg). Roentgen, in 1896, had demonstrated the existence of X-rays (electromagnetic radiation) emitted when fast electrons strike matter and, initially, it was thought that the first discovered of the new 'atomic' radiations (by Becquerel in 1896) was a form of X-ray. Becquerel had found that uranium compounds emitted a form of radiation (now known as β-radiation) spontaneously but that, unlike

X-rays, it could be deflected by a magnetic field implying that the β-rays were streams of charged particles.

Further work by Pierre and Marie Curie and by Rutherford in 1897–1898 established the existence of another form of charged-particle radiation (now referred to as α-radiation) which was much less penetrating than β-radiation. Typically, α-rays could hardly penetrate $\simeq 10^{-4}$ m of lead compared with $\simeq 10^{-3}$ m for β-rays.

Finally, in 1900, Villard established the existence of an even more penetrating form of radiation (now referred to as γ-radiation) which remained undeflected by a magnetic field. This radiation is now known to be electromagnetic in nature (like X-rays) but of higher penetrating power (typically $\simeq 10^{-1}$ m of lead).

These three forms of radiation were investigated intensively during the first decade or so of the 20th century, particularly by Rutherford, and by studying the deflection of α- and β-rays in magnetic and electric fields values for the charges and masses for α- and β-particles were obtained, namely

$$e_\alpha \simeq +3.2 \times 10^{-19}\,\text{C} \qquad m_\alpha \simeq 6.6 \times 10^{-27}\,\text{kg}$$
$$e_\beta \simeq -1.6 \times 10^{-19}\,\text{C} \qquad m_\beta \simeq 9.1 \times 10^{-31}\,\text{kg}$$

These values are in agreement with the α-particles' being positively charged helium atoms with a charge equal to twice the magnitude of the electron charge, and the β-particles' being electrons. It also emerged that whilst the α-particles were emitted with definite energies the β-particles (electrons) had a continuous spread of energies which, incidentally, were much higher (of the order MeV) than the energies then obtained for electrons in discharge tubes. Such high energies, which are commensurate with the mass energy of an electron (0.51 MeV – section 1.1), mean that any theoretical treatment of β-radiation must use relativistic mechanics (Chapter 6).

The foregoing results were obtained after much detailed experimental work. Usually the radiations were detected using ionization chambers in which the radiation causes a current, which can be measured, to flow through a gas contained in the chamber.

1.3 ATOMIC STRUCTURE AND THE NUCLEUS

The first attempt to formulate a model of the atom was by J. J. Thomson in 1898. Given the experimental information then available it seemed clear that atoms must contain electrons. However, atoms are electrically neutral and so it was a natural next step for him to suppose that the electrons were embedded in positively charged matter which was also responsible for the bulk of the mass of an atom (remember that atomic masses are measured in units of 1.66×10^{-27} kg whilst the mass of an electron is only 9.1×10^{-31} kg). so emerged the Thomson 'plum-pudding' model in which the atom is envisaged as a sphere of positively charged material whose radius is of the order 10^{-10} m and whose charge is exactly cancelled by the negative charges of the electrons. This model we now know to be utterly wrong but it was not until 1911 that Geiger and Marsden, at the suggestion of Rutherford, established the currently accepted model of an atom.

This classic experiment involved a study of the way in which α-particles are scattered by gold atoms owing to the interaction of their electric charges. They used a radium source emitting α-particles and produced a collimated beam of these particles (Fig. 1.1) which were then scattered by a thin ($\simeq 10^{-5}$ m thick) gold foil. The scattered α-particles were detected by microscopic examination of the scintillations produced when they struck a screen of zinc sulphide.

If the Thomson model was correct then it was expected that virtually all the α-particles would go through the screen and at most suffer a very small angular deviation from their initial path. This expectation was based on the fact that in the Thomson model there is no great concentration of electric charge and so the resultant electric field experienced by the α-particle as it passed through the atom

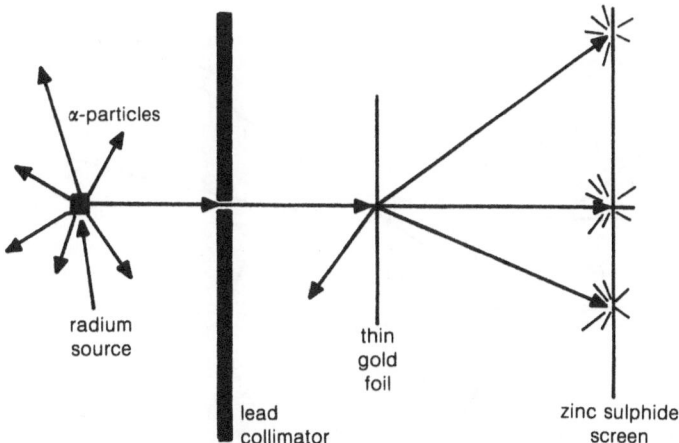

Fig. 1.1 Schematic representation of the Geiger–Marsden scattering experiment.

would be small. However, even though not expected, Rutherford suggested that they should look for large angle scattering. This they did and were astounded to find that some α-particles were scattered through very large angles, even greater than 90°. Rutherford later recorded his great surprise at this outcome: 'It was quite the most incredible event that has ever happened to me in my life. It was almost as incredible as if you had fired a 15-inch shell at a piece of tissue paper and it came back and hit you.'

To account for these experimental results, Rutherford proposed that all the positive charge of an atom was concentrated at its centre whilst the electrons occupied a sphere of atomic dimensions ($\simeq 10^{-10}$ m). Assuming that the positive charge was attached to some form of matter, it followed that virtually all the mass of an atom was also concentrated at the centre. This meant that the main agent in deflecting an α-particle when it collided with an atom would be this high concentration of charge and mass at the centre – the **atomic nucleus** – and that multiple scattering by the electrons could be ignored. On this basis it was easy to understand the qualitative outcome of the Geiger–Marsden experiment. Most α-particles moving through the outer spaces of an atom would suffer little deflection, but those few which came close to the nucleus would experience an intense repulsive electric field and be deflected through large angles (Fig. 1.2). The extreme situation would be a 'head-on' collision when

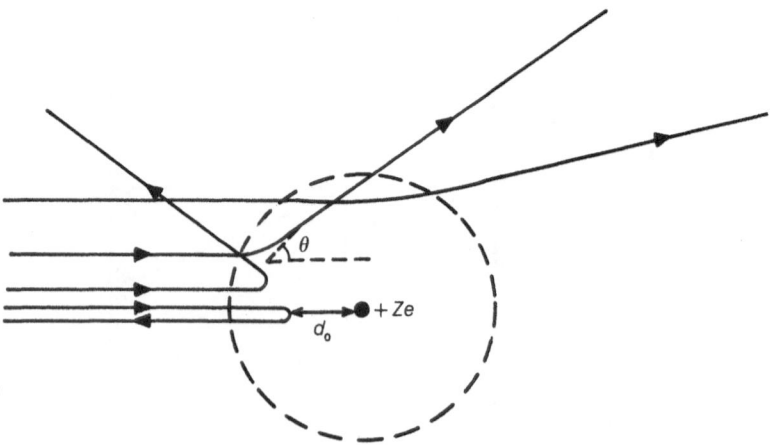

Fig. 1.2 Rutherford scattering of α-particles.

the α-particle would be turned around in its tracks (180° scattering).

Of course, it was necessary to quantify the foregoing discussion and to test the theory by very careful experiment. This was done by Rutherford, Geiger and Marsden during the period 1911–1913. Supposing that there are Z electrons in an atom it follows that their total charge is $- Ze$, where e is the magnitude of the electronic charge. For the atom to be neutral this then requires the nuclear charge to be $+ Ze$. The nearest an α-particle could get to the nucleus would be in a 'head-on' collision and it is interesting to calculate this distance, d_0.

Suppose that the mass of the α-particle is M and its incident speed is v. Assuming for the moment that the nuclear charge is concentrated at a point, d_0 can be obtained by equating the initial kinetic energy of the α-particle ($\frac{1}{2}Mv^2$) to the potential energy ($2Ze^2/4\pi\varepsilon_0 d_0$) it will have acquired when it momentarily comes to rest before its motion is reversed. This gives

$$d_0 = \frac{1}{4\pi\varepsilon_0} \frac{4Ze^2}{Mv^2}$$

Typically α-particles in the Geiger–Marsden experiment had energies of the order 7 MeV (i.e. $\frac{1}{2}Mv^2 = 7 \times 1.6 \times 10^{-13}$ J) and, for gold, $Z = 79$. Substituting in the above expression then gives $d_0 \simeq 3 \times 10^{-14}$ m. Rutherford went further and calculated the probability of an α-particle's being scattered through an angle θ and found that this probability for a given foil was proportional to $d_0^2 \, \mathrm{cosec}^4 \frac{1}{2}\theta$ (section 5.5). Experiment confirmed this prediction and therefore supported in particular the assumption that the nucleus behaved as though its charge was concentrated at a point. Of course, in these experiments, as has just been shown, the nearest distance of approach to the nucleus is $\simeq 3 \times 10^{-14}$ m and so the correct conclusion to draw is that the size of a gold nucleus must certainly be less than 3×10^{-14} m or 30 fm where 1 fm = 1 Fermi (or 1 femtometre) = 10^{-15} m. Subsequent experiments over the years using much more energetic particles have shown significant deviations from Rutherford scattering which have enabled the precise sizes of nuclei to be determined. Even so, the Geiger–Marsden experiments established unequivocally the nuclear model of the atom and that nuclear radii were at least some four orders of magnitude smaller than atomic radii.

Following on from this was the massive advance of Bohr in 1913 in formulating his theory of electrons orbiting the nucleus in their different quantum states and the full development of quantum theory in the middle 1920s.

1.4 NUCLEAR CONSTITUENTS

Although the nuclear atomic model was firmly established in 1911–1913 it was another 20 years before the basic constituents of the nucleus were confidently identified. The work of Rutherford, Geiger, Marsden and others showed that an atomic nucleus was relatively very small ($\simeq 10^{-14}$ m) and was characterized by its charge ($+Ze$) where Z equalled the number of electrons in the corresponding neutral atom and its mass M which was frequently close in value to an integer number of atomic mass units. For heavy nuclei this integer was of the order of twice the value of Z.

Bearing this in mind, one possible structure for a nucleus was to suppose that it had two types of component – hydrogen nuclei (now called **protons**) and **electrons**. Since a proton has a mass close to 1 u (section 1.1) a suitable number would ensure that the nucleus had approximately the right mass. The resultant positive charge would, however, be too large and hence the inclusion of the electrons to ensure the correct total charge. However, it soon became clear that this model was untenable. The development of quantum mechanics brought with it Heisenberg's uncertainty relation

$$\Delta p\, \Delta x \geqslant \frac{\hbar}{2}$$

where Δp is the uncertainty in the momentum of a particle whose position is localized by Δx and $\hbar$ is Planck's quantum constant h divided by 2π ($\hbar = 1.054 \times 10^{-34}$ J s). For an electron localized in, for example, a nucleus of radius $\simeq 10^{-14}$ m ($\simeq \Delta x$) the corresponding value of Δp is $\simeq 5 \times 10^{-21}$ kg m s^{-1}. In turn, this means that the kinetic energy of the electron is about 10 MeV (note: this needs to be calculated relativistically and is simply approximated by pc since it is so much larger than the electron rest mass energy, 0.51 MeV). Electrical forces could not possibly confine such high energy electrons in the nucleus and if additional forces existed they would be expected to influence the orbiting atomic electrons. Later other nuclear data (e.g. nuclear spins, nuclear magnetic moments) were shown to be at complete variance with the concept of electrons as nuclear constituents and, in any case, in 1932 it was shown by Chadwick that nuclei contained **neutrons** (denoted by n).

A neutron is an electrically neutral particle having very nearly the same mass as a proton:

$$m_{\text{proton}} = m_{\text{p}} = 1.007\,28\ \text{u}, \quad m_{\text{neutron}} = m_{\text{n}} = 1.008\,67\ \text{u}$$

This meant that any given nucleus would contain the appropriate number of protons to give the correct positive charge together with sufficient neutrons to give the correct mass.

The possible existence of some sort of neutral particle had been mooted by Rutherford several years before its discovery but, being neutral, it was difficult to detect. From earlier work it was known that when beryllium was bombarded with α-particles from a radioactive source a highly penetrating radiation was emitted. It was known that this radiation could knock protons out of hydrogen-rich substances such as paraffin. At first it was thought that the radiation was some form of high energy γ-ray, but Chadwick suggested that the radiation consisted of neutral particles. He made a detailed study of the energetics of the process (Fig. 1.3), detecting the protons knocked out of the paraffin using a Geiger counter. The latter is a device in which a massive electrical discharge is triggered by protons ionizing a low pressure gas contained in a tube having electrodes maintained at a high potential difference. His experiments established not only the existence of neutrons but also that the neutron mass was very close to that of the proton.

The present-day picture of the nucleus thus emerged in which there are two types of nuclear constituent – protons and neutrons. Since the masses of protons and neutrons are very close and since their nuclear (as distinct from their electromagnetic) properties are virtually identical, protons and neutrons are referred to collectively as **nucleons**. A specific nucleus can thus be specified in terms of the following:

Z = number of protons
   = atomic number
N = number of neutrons
A = total number of neutrons and protons
   = number of nucleons
   = mass number
   = $Z + N$

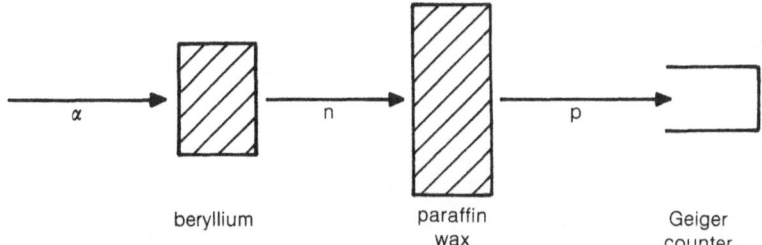

Fig. 1.3 Schematic representation of the Chadwick neutron experiment.

A specific nuclear species is denoted by the symbolism $^{A}_{Z}X$ where X is the usual chemical symbol for the element referred to, e.g. $^{4}_{2}He$, $^{19}_{9}F$, $^{208}_{82}Pb$. Chadwick's experiment can thus be represented symbolically by

$$^{9}_{4}Be + ^{4}_{2}He \rightarrow ^{12}_{6}C + n$$

Since Z determines the nuclear charge and hence the number of electrons and hence the chemical properties of an element, it is easy to understand the existence of **isotopes** – atoms having the same chemical properties but different masses. They are simply atoms whose nuclei contain the same number of protons but different numbers of neutrons, e.g. $^{12}_{6}C$ and $^{13}_{6}C$.

2

General properties of the nucleus

Having described the way in which the nuclear model of the atom came into being we now consider some of the general properties of nuclei in terms of which they can be characterized. A historical approach will no longer be used and the descriptions and specifications of the different properties will be as up to date as possible. We start from the basic conclusion of the last chapter that an atomic nucleus is an assembly of Z protons and $N(=A-Z)$ neutrons confined to a region whose linear dimension is of the order of magnitude of 10 fm or less. A key issue, of course, is how this assembly is held together. Clearly, in order to achieve this, there must be a powerful attractive force between the nucleons – the nuclear force. This will be investigated in detail in Chapter 3 and meanwhile it will simply be assumed that it is present.

2.1 NUCLEAR SIZES

The Rutherford–Geiger–Marsden experiment and its analysis set the pattern for most future studies of nuclear sizes. Basically the approach is to scatter charged particles by nuclei and to measure deviations from the predictions for the scattering pattern assuming a point nucleus. Such deviations can give information not only about the actual size of a nucleus but also about the way the nuclear charge is distributed throughout the nucleus. The scattering pattern has to be evaluated quantum mechanically taking into account the wave-like nature exhibited by particles. The incoming wave will have the form $\exp(2\pi i x/\lambda)$ where λ is the de Broglie wavelength ($\lambda = h/p$, $p = $ momentum) for the particles and the process is essentially one of diffraction of this wave by the nucleus. To obtain detailed information about the charge distribution requires that $\lambda/2\pi$ is significantly smaller than the nuclear size, i.e. $\simeq 1$ fm and even less for lighter, smaller nuclei.

High energy electrons are particularly useful for studying nuclear sizes since only electromagnetic forces come into play; this is not the case with neutrons, protons, α-particles etc. which also experience the nuclear force and for which the analysis is more complicated. But it must be recognized that using electrons only gives information about the nuclear charge (or proton) distribution. For electrons, taking $\lambda/2\pi$ to be 1 fm, the de Broglie relationship gives $p \simeq 1 \times 10^{-19}$ kg m s^{-1}. This is a very high momentum and the electron motion is relativistic. We can therefore use the approximate expression $E = pc$ to calculate the corresponding energy. This gives $E \simeq 200$ MeV. Higher energies than this are needed for light nuclei. Also, since the energies are relativistic, Rutherford's expression for scattering from a point charge cannot be used and deviations from its relativistic version (due to Mott) must be studied.

Many electron scattering experiments on the full spectrum of stable nuclei have been carried out using electrons accelerated to energies in the hundreds of MeV region (section 5.3 includes a brief discussion of accelerators). The outcome of these experiments is that a typical nuclear charge distribution is found to have an approximate shape of the kind shown in Fig. 2.1. This distribution is described in terms of three quantities.

1. The **charge density** ρ_{ch} is normalized so that when integrated over the nuclear volume it gives the total charge Z measured in units

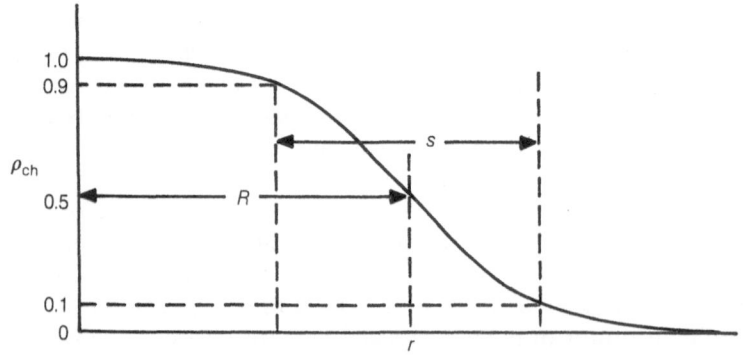

Fig. 2.1 Nuclear charge density distribution.

of the proton charge, i.e.

$$4\pi \int \rho_{ch}(r)r^2 \, dr = Z \tag{2.1}$$

2. The **'half' radius** R is the value of the radius at which the charge density has decreased to half its central value and is loosely referred to as the charge radius. Typically R is given by

$$R \simeq r_0 A^{1/3} \, \text{fm} \tag{2.2}$$

where $r_0 \simeq 1.1 \, \text{fm}$.

3. The **surface 'width'** s measures the radial distance over which the charge density reduces from 90% to 10% of its central value. This has much the same value for all nuclei:

$$s \simeq 2.5 \, \text{fm} \tag{2.3}$$

These results mean that nuclear charge radii, R, vary from around 2 fm for a light nucleus to around 7 fm for a very heavy nucleus. It also follows from eq. (2.2) that the volume of a nucleus ($\propto R^3$) is proportional to A, i.e. to the number of nucleons it contains. Further, the value for s (eq. (2.3)) means that very light nuclei are virtually all surface with no region of constant density at the centre.

As has already been pointed out, nuclear scattering of electrons only gives information about the distribution of protons. To study the matter (i.e. proton and neutron) distribution requires more complicated analysis of the scattering of, for example, α-particles or neutrons, which interact through the nuclear force with both protons and neutrons in the nucleus. Such analysis indicates that the matter distribution, as might be expected, has a similar shape to the charge distribution although there are indications that the neutron distribution may extend a little (by $\simeq 0.1 \, \text{fm}$) beyond the proton distribution.

It is interesting to consider the difference between the density of nuclear matter and ordinary matter. For a nucleus containing A nucleons the mass is of the order Am_p where m_p ($\simeq m_n$) is the mass of a proton ($\simeq 1.67 \times 10^{-27} \, \text{kg}$), and the volume is $4\pi R^3/3$ where $R \simeq 1.1 A^{1/3} \, \text{fm}$. The density (mass/volume) is clearly independent of A and is approximately $3 \times 10^{17} \, \text{kg m}^{-3}$. This is to be compared with the density of ordinary matter which is typically $\simeq 10^3 \, \text{kg m}^{-3}$! This is why something as heavy as a star can evolve into a neutron star, consisting simply of nuclear matter, and having a radius of only a few km.

Finally, at this same level of approximation, we estimate the average separation between nucleons in a nucleus. If this separation is denoted by α, the average volume 'occupied' by each nucleon will be $\simeq \alpha^3$, so that the total nuclear volume is $\simeq A\alpha^3$. Comparing this with the expression for the volume given in the last paragraph then gives $\alpha \simeq 1.6r_0 \simeq 1.8\,\text{fm}$.

2.2 NUCLEAR MASS AND BINDING ENERGY

The mass of a nucleus with atomic number Z and mass number A is clearly related to the sum of the masses of its constituent nucleons, i.e. $Zm_p + Nm_n$ where m_p and m_n are the masses of the proton and neutron respectively and $N = A - Z$ (section 1.4). However, it must be remembered that the nucleons are bound together by the nuclear force so that energy is required to remove one or more of them if the nucleus is stable. Conversely, if the nucleus was to be assembled by bringing together its constituent nucleons from infinity, energy (denoted by B) would be released. Given the Einstein relation between mass and energy ($E = mc^2$) and the conservation of energy, it therefore follows that the mass of a stable nucleus is somewhat less than the total mass of its constituents by an amount B/c^2. Thus, if M_{nuc} is the mass of a nucleus, we have

$$M_{\text{nuc}} = Zm_p + Nm_n - \frac{B}{c^2} \qquad (2.4)$$

or

$$B = (Zm_p + Nm_n - M_{\text{nuc}})c^2 \qquad (2.5)$$

Alternatively, the above relations can be expressed in terms of atomic masses, rather than nuclear, bearing in mind that the mass of an atom, M_{at}, characterized by Z and A is given by

$$M_{\text{at}} = M_{\text{nuc}} + Zm_e - \frac{B_{\text{at}}}{c^2} \qquad (2.6)$$

where B_{at} is the binding energy of the Z atomic electrons in the atom and is generally neglected (typically $B_{\text{at}}/c^2 \sim 10^{-6}M_{\text{at}}$). Thus eq. (2.6) can be written

$$B = (ZM_H + Nm_n - M_{\text{at}})c^2 \qquad (2.7)$$

where M_H is the mass of the hydrogen atom ($\simeq m_p + m_e$) and the masses of Z electrons implicitly included in the first and third terms clearly cancel.

Table 2.1 Nuclear masses and binding energies

Nucleus	$M(A, Z)$ in u	$B(A, Z)$ in MeV	$B(A, Z)/A$ in MeV
$^{4}_{2}\text{He}$	4.002 603 3	28.3	7.07
$^{12}_{6}\text{C}$	12.000 000 0	92.2	7.68
$^{133}_{54}\text{Xe}$	132.905 82	1119.0	8.41
$^{238}_{92}\text{U}$	238.050 76	1803.0	7.58

The terms nuclear mass and atomic mass are loosely used interchangeably although they clearly differ by the mass of Z electrons. For our purposes the difference (of order 0.1%, or less) is generally unimportant. When precise masses are needed we shall use the notation M or $M(A, Z)$ to denote the **atomic** mass. The nuclear binding energy will similarly be denoted by B or $B(A, Z)$.

Nuclear masses are measured using **mass spectrometers** in which the motion of singly ionized atoms moving through crossed electric and magnetic fields is carefully measured. A few typical masses and corresponding binding energies are shown in Table 2.1. Included in the table are values of $B(A, Z)/A$ which is an expression for the average binding energy per nucleon, and it will be noted that this has roughly the same value for the different nuclei. This shows itself even more strongly if the value of $B(A, Z)/A$ is plotted as a function of A for all stable nuclei as in Fig. 2.2. From the resultant curve the following features emerge.

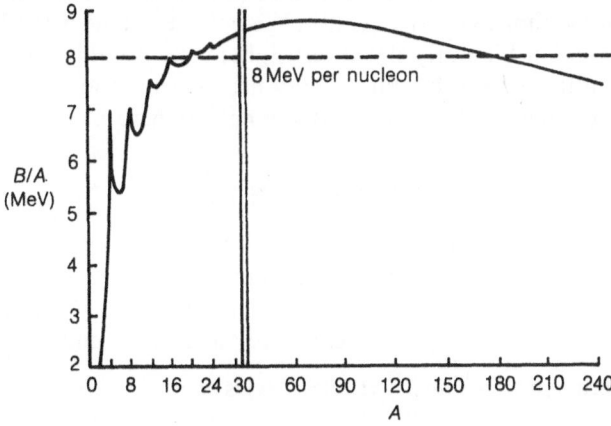

Fig. 2.2 Binding energy per nucleon (B/A) as a function of mass number A.

1. Beyond $A \simeq 12$, B/A generally lies between about 7.5 and 8.5 MeV with an average value of around 8 MeV.
2. For very light nuclei B/A is significantly lower but with pronounced peaks for $A = 4$ (He), 8 (Be), 12 (C), etc.
3. The maximum value of B/A is around 8.8 MeV in the region $A \simeq 55$–60, e.g. ^{56}Fe, for which $B/A = 8.79$ MeV.
4. Beyond this maximum the value of B/A gradually decreases.

All the foregoing features indicate various properties of the nuclear force and these will be explored in Chapter 3. Suffice it here to make one or two general remarks.

First, although B/A is the average binding energy per nucleon it is not quite the same as the energy needed to remove a particular neutron or proton from the nucleus. This is given by the **separation** energy which is obtained as follows. Suppose a neutron is removed from a nucleus (A, Z) with mass $M(A, Z)$. There would remain a nucleus $(A - 1, Z)$ and a neutron, with total mass $M(A - 1, Z) + m_n$, so that the amount of energy needed to separate the neutron from the original nucleus, which is denoted by $S_n(A, Z)$, is simply

$$S_n(A, Z) = [M(A - 1, Z) + m_n - M(A, Z)]c^2$$

Similar expressions can be written down for the proton, α-particle etc. separation energies.

Second, inspection of the binding energy curve in Fig. 2.2 shows that if a very heavy nucleus (mass number A) is separated into two roughly equal nuclei with mass numbers $\simeq A/2$ then, since the binding energy per nucleon has increased by about 1 MeV, it follows that an amount of energy $\simeq A$ MeV is released. This is the basis of **nuclear fission**. Similarly, if two very light nuclei are combined there is again an increase in binding energy per nucleon and a consequent release of energy. This is the basis of **nuclear fusion**.

2.3 NUCLEAR STABILITY

In section 2.2 reference has been made from time to time to **stable** nuclei and this term must now be carefully explained. A nucleus (A, Z) is unstable if it eventually disintegrates in some way. There are many ways in which disintegration can take place, for example:

1. **α-decay**, in which an α-particle ($^4_2$He) is emitted leaving the nucleus $(A - 4, Z - 2)$;
2. **β^--decay**, in which a neutron changes into a proton with the

emission of an electron and an antineutrino (Chapter 6) leaving the nucleus $(A, Z + 1)$;

3. **nucleon emission,** in which a neutron or a proton is emitted leaving the nuclei $(A - 1, Z)$ or $(A - 1, Z - 1)$ respectively;
4. **fission,** in which the nucleus splits into two roughly equal parts together with two or three neutrons.

Given the conservation of mass–energy it is clear that such disintegrations will only be possible if the mass of the initial nucleus is larger than the sum of the masses of the disintegration products, any excess energy being carried away as kinetic energy. Thus, for α-decay to take place, for example, we must have

$$M(A, Z) > M(A - 4, Z - 2) + M(4, 2)$$

Conversely a nucleus is completely stable if its mass if less than the sum of the masses of all possible disintegration products, i.e.

$$M(A, Z) < \sum_i M_i$$

where M_i is the mass of a disintegration product.

The time for which an unstable nucleus can exist before it disintegrates can vary enormously through from geological times to times as short as 10^{-23}s depending on the energetics and the nature of the disintegration process. These matters will be discussed in detail later on (e.g. Chapters 5 and 6). Frequently nuclei which, although unstable, exist for a relatively long time are referred to as **metastable**. A stable nucleus, of course, has an infinite lifetime.

It is interesting to see how the numbers of protons and neutrons in a nucleus correlate with nuclear stability. Consider first the number of stable nuclei in relation to the evenness or oddness of Z and N shown in Table 2.2.

Table 2.2 Numbers of stable nuclei

A	Z	N	Number of stable nuclei
even	even	even	165
odd	odd	even	50
odd	even	odd	55
even	odd	odd	4

The distribution indicates unequivocally that stability is closely associated with evenness in the number of protons and neutrons and suggests that the nuclear force must be particularly effective when neutrons and protons are 'paired off'. The similarity in number of 'odd–even' and 'even–odd' nuclei suggests that neutrons and protons behave in much the same way as far as the nuclear force is concerned.

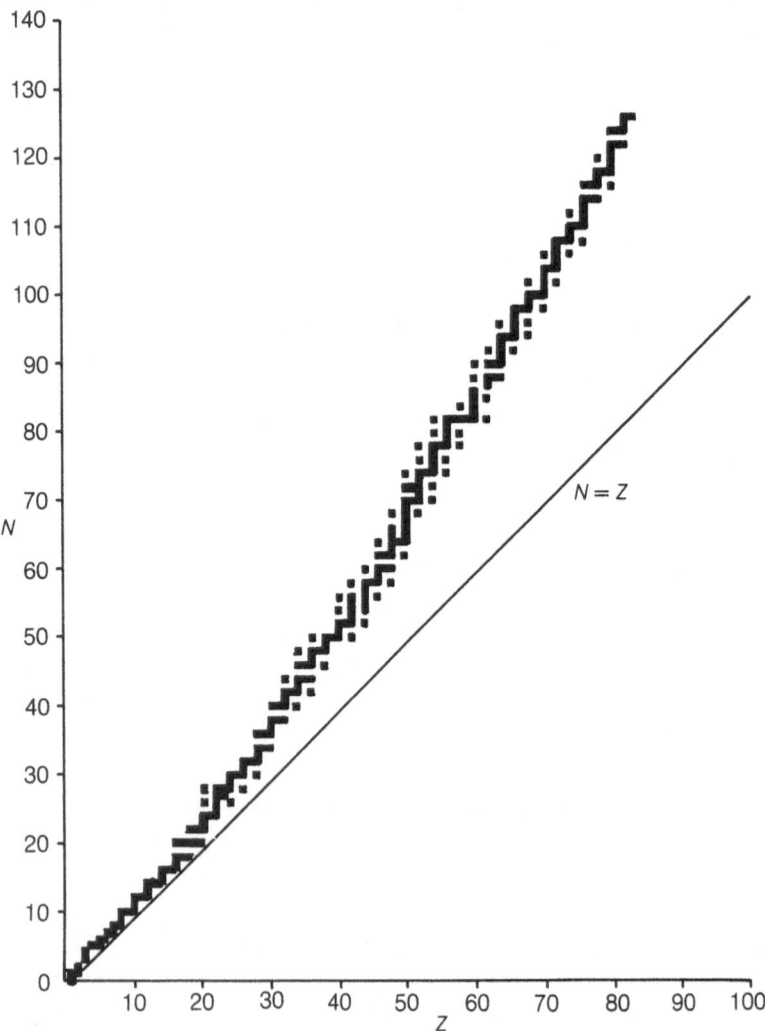

Fig. 2.3 Segre chart of stable nuclei as a function of N and Z.

Finally, the paucity in number of 'odd–odd' nuclei (restricted to very light nuclei such as $^6_3$Li and $^{10}_5$B) is particularly striking.

Another way of looking at stability is to enter each stable nucleus as a point on a plot of N against Z as in Fig. 2.3 (known as a Segre chart). Here it will be noted that the 'valley of stability', as it is frequently called, veers away from the line $N = Z$ in the direction of proportionately larger N. This can be simply understood in qualitative terms as a result of the Coulomb repulsion which exists between protons. This repulsion works against the attractive nuclear force which is responsible for nuclear stability and hence it becomes energetically favourable to have proportionately fewer protons in stable nuclei. The extent to which this happens will be dealt with quantitatively in Chapter 4.

2.4 GROUND AND EXCITED STATES OF NUCLEI

A nucleus is a complicated structure of A interacting nucleons and in principle, and to some degree in practice (Chapter 4), an appropriate quantum mechanical Hamiltonian H can be written down in terms of the nuclear and Coulomb interactions. This Hamiltonian will have a series of eigenfunctions ψ_n with corresponding energy eigenvalues E_n given by the Schrödinger equation, i.e.

$$H\psi_n = E_n\psi_n \tag{2.8}$$

The symbol n signifies all the quantum numbers necessary to define the state of the nucleus and these will be specified in more detail later.

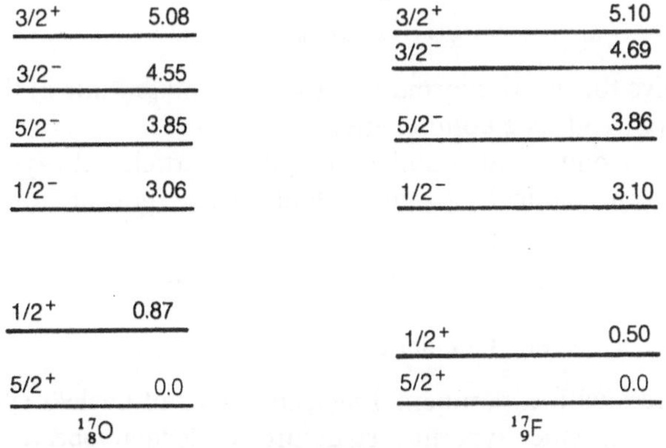

$3/2^+$	5.08		$3/2^+$	5.10
			$3/2^-$	4.69
$3/2^-$	4.55			
$5/2^-$	3.85		$5/2^-$	3.86
$1/2^-$	3.06		$1/2^-$	3.10
$1/2^+$	0.87			
			$1/2^+$	0.50
$5/2^+$	0.0		$5/2^+$	0.0
$^{17}_8$O			$^{17}_9$F	

Fig. 2.4 Low lying energy levels of $^{17}_8$O and $^{17}_9$F.

The state of lowest energy is known as the nuclear **ground state** and the remainder as **excited states**. Nuclear energy level diagrams (Fig. 2.4 provides an example for the nuclei ^{17}O and ^{17}F) give the energies of these excited states which are shown at the right of each energy level (in MeV). It is also conventional to label a state with its spin quantum number and its parity and these are introduced in sections 2.5 and 2.6.

2.5 NUCLEAR SPIN

Protons and neutrons, the constituents of nuclei, both have spin quantum number $\frac{1}{2}$ (i.e. $s_p = s_n = \frac{1}{2}$). In quantum mechanics the spin of the proton is represented by a vector operator s_p such that the eigenvalue of s_p^2 is $\frac{1}{2}(\frac{1}{2} + 1)\hbar^2$ and of s_{pz} is $+\frac{1}{2}\hbar$ or $-\frac{1}{2}\hbar$ and similarly for s_n. In addition, the nucleons may also have orbital angular momentum by virtue of their motion in the nucleus; this is represented by an angular momentum quantum number $l(= 0, 1, 2, \ldots)$ for each nucleon.

The sum total of the spin and orbital angular momenta of the nucleons, the total intrinsic angular momentum of the nucleus, is referred to as the **nuclear spin** and the associated quantum number is denoted by J. As with the spins of individual nucleons, there is a corresponding nuclear spin operator J. If the quantum mechanical state of a nucleus is represented by a wave function ψ_{JM} (where $M = J, J - 1, \ldots, -J$) then the main properties of J are represented by the two eigenfunction equations:

$$J^2\psi_{JM} = J(J + 1)\hbar^2\psi_{JM} \tag{2.9}$$

$$J_z\psi_{JM} = M\hbar\psi_{JM} \tag{2.10}$$

which give the usual information about the magnitude of the square of the spin and its z components.

Since an odd (even) number of spin $\frac{1}{2}$ particles always combine quantum mechanically to give a half-integer (integer) total spin, it follows that

odd A nuclei have $J = 1/2, 3/2, 5/2, \ldots$

even A nuclei have $J = 0, 1, 2, \ldots$

This agrees with experimental measurements of nuclear spins using studies of atomic hyperfine structure, molecular spectra, nuclear magnetic resonance and other techniques. In addition it is found that for 'even–even' (Z even, N even) nuclei the nuclear ground state spin is always $J = 0$. In Fig. 2.4 the value of the spin where it is known is placed at the left of each energy level.

2.6 NUCLEAR PARITY

The wave function for a nuclear energy eigenstate just discussed is a function of the coordinates $r_1, r_2, \ldots, r_A$ of the A nucleons which it contains and, in fuller detail, should be written $\psi(r_1, r_2, \ldots, r_A)$. As well as being characterized by a definite spin quantum number (J) it is also characterized by being either even or odd under reflection of coordinate axes. This can be expressed simply as follows.

We define a **parity** operator P which has the property of reflecting each coordinate r_i through the origin ($r_i \rightarrow -r_i$), i.e.

$$P\psi(r_1, r_2, \ldots, r_A) = \psi(-r_1, -r_2, \ldots, -r_A) \qquad (2.11)$$

The evenness or oddness of ψ is then manifest in the following eigenvalue equation satisfied by all nuclear energy eigenstates:

$$P\psi(r_1, r_2, \ldots, r_A) = P\psi(r_1, r_2, \ldots, r_A) \qquad (2.12)$$

where $P = \pm 1$. A state with $P = +1 (-1)$ is said to have **even (odd)** parity. This means that a nuclear state can be labelled with both its spin and its parity, the usual symbolism being J^P (e.g. $1^-, 2^+, \frac{1}{2}^-$) – see Fig. 2.4. In particular, the ground states of even–even nuclei are invariably found to be 0^+. (In the foregoing discussion it has been implicitly assumed that the neutron and proton have the same intrinsic parity. This is a convention and the point is further discussed in section 7.3.)

Further, so long as the nuclear Hamiltonian satisfies

$$H(r_1, r_2, \ldots, r_A) = H(-r_1, -r_2, \ldots, -r_A)$$
$$= PH(r_1, r_2, \ldots, r_A) \qquad (2.13)$$

which holds with considerable accuracy, then the parity operator P commutes with H, i.e.

$$[P, H] = 0 \qquad (2.14)$$

In turn this means that parity is a constant of the motion; parity is **conserved** in nuclear processes. (We shall see later on in section 9.2 that there is a weak nuclear force ($\simeq 10^{-6}$–10^{-7} times as strong as the strong nuclear force) which changes sign on reflection of axes so invalidating eq. (2.13). However, because it is so weak, its parity non-

conserving effects are very small.) The determination of the parities of
nuclear states is achieved by studying the spatial distribution of
particles and photons in nuclear processes, particularly in β- and γ-
decay (Chapter 6).

2.7 NUCLEAR ELECTROMAGNETIC MOMENTS

The static electromagnetic properties of nuclei are specified in terms
of electromagnetic moments which give information about the way in
which magnetism and charge is distributed throughout the nucleus.
The two most important moments are the **magnetic moment** (μ) and
the **electric quadrupole moment** Q.

2.7.1 Nuclear Magnetic Moments

For an electron of charge $-e$ and mass m orbiting with angular
momentum L, it is well known that there is an associated orbital
magnetic moment given by

$$\mu_L = -\frac{e}{2m}L$$

Similarly, associated with its spin s there is an intrinsic magnetic
moment

$$\mu_S = -g\frac{e}{2m}s \tag{2.15}$$

where e and m are the charge and mass of the electron respectively. g
is known as the g-factor and Dirac's relativistic wave equation gives
precisely $g = 2.0000\ldots$. However, complicated quantum electro-
dynamical effects modify this value slightly to $g = 2.002\,319\,2\ldots$.

The actual value of the electron magnetic moment, μ, is defined as
the eigenvalue of μ_{Sz} when the electron is in the spin substate $m_s = \frac{1}{2}$.
Clearly

$$\mu_{Sz} = -g\frac{e}{2m}s_z$$

where s_z has eigenvalues $m_s\hbar$ with $m_s = \pm\frac{1}{2}$. Thus, for $m_s = \frac{1}{2}$ we have
for the value of the magnetic moment

$$\mu_S = -\frac{1}{2}g\frac{e\hbar}{2m} = -\frac{1}{2}g\mu_B \tag{2.16}$$

where $\mu_B (= e\hbar/2m)$ is known as the Bohr magneton and has the value $\mu_B = 9.274... \times 10^{-24} \, \text{J T}^{-1}$.

All elementary particles with spin (apart from the neutrino) have an intrinsic magnetic moment and exactly analogous relationships hold. Thus, the magnetic moment operators for the proton and the neutron are defined as

$$\mu_p = g_p \frac{e}{2m_p} s_p$$

$$\mu_n = g_n \frac{e}{2m_p} s_n \tag{2.17}$$

and their values are now characterized by the mass of the proton (m_p) and different g-factors for each particle. In turn (cf. eq. (2.16)) the corresponding values of their magnetic moments are given by

$$\mu_p = \tfrac{1}{2} g_p \mu_N$$

$$\mu_n = \tfrac{1}{2} g_n \mu_N \tag{2.18}$$

where, analogous to the Bohr magneton μ_B, the nuclear magneton μ_N is defined as

$$\mu_N = \frac{e\hbar}{2m_p} = 5.050\,78... \times 10^{-27} \, \text{J T}^{-1} \tag{2.19}$$

The actual measured nucleon g-factors are

$$g_p = +5.5856... \quad \text{and} \quad g_n = -3.8262...$$

and so the corresponding magnetic moments are

$$\mu_p = +2.7928...\mu_N \quad \text{and} \quad \mu_n = -1.9131...\mu_N$$

Because of the dependence of the values for μ on the mass of the proton, it follows that nucleon magnetic moments are some three orders of magnitude smaller than the magnetic moment of the electron. In addition, they differ significantly from the value predicted for a spin $\tfrac{1}{2}$ particle by the Dirac equation ($\mu_p = 1\mu_N$, $\mu_n = 0$). This is attributed to the fact, whose details will emerge in Chapter 8, that unlike the electron, for which μ is very close to the Dirac value, nucleons are finite in size and have a complicated internal structure.

Turning now to atomic nuclei themselves, the intrinsic magnetic moments of the constituent protons and neutrons will contribute to a total magnetic moment and there will be a further contribution from any orbital motion of the (charged) protons. As with particles the total magnetic moment operator $\boldsymbol{\mu}_J$ and the corresponding magnetic moment μ_J are again simply related to the spin operator $\boldsymbol{J}$ and the value of the spin, J, by completely analogous relations to those exhibited in eqs (2.15)–(2.18), i.e.

$$\boldsymbol{\mu}_J = g_J \frac{e}{2m_p} \boldsymbol{J} \quad \text{and} \quad \mu_J = g_J \mu_N J \tag{2.20}$$

where g_J is the nuclear g-factor.

Magnetic moments of most nuclear ground states as well as many excited states have been measured. The techniques used include hyperfine structure studies, microwave spectroscopy, nuclear magnetic resonance, paramagnetic resonance, use of molecular and atomic beams and nuclear alignment. It is found that nuclear moments lie in the approximate range from $-2\mu_N$ to $+6\mu_N$. Their values are a useful test (section 4.2.3) of detailed theories of nuclear structure.

2.7.2 Nuclear Electric Quadrupole Moments

Discussion about the nucleus so far has implicitly assumed that it has spherical symmetry. For example, the charge density ρ_{ch} (section 2.1) was taken to be a function of the radial coordinate r only, implying that surfaces of constant ρ are spherical. In other words, the nucleus has been taken to have the form of a minute fuzzy ball. Some nuclei do indeed have spherical symmetry, but many do not and the quantity which most simply measures the deviation from spherical symmetry is known as the electric quadrupole moment and is denoted by the symbol Q.

The formal classical definition of Q (denoted by Q_0) is as follows:

$$Q_0 = \int \rho_{ch}(r)(3z^2 - r^2) \, dV \tag{2.21}$$

where the z-axis is taken to be along the axis of symmetry defined by the nuclear spin. The charge density is now shown to be dependent on the vector r, i.e. it can vary not only with the value of r but also with its direction.

The above expression is simply the average of $3z^2 - r^2$ taken over the charge density distribution, i.e. it could be written

$$Q_0 = Z(3\langle z^2 \rangle - \langle r^2 \rangle) \tag{2.22}$$

where Z is the total nuclear charge measured in units of e. Q_0 thus has the dimensions (length)^2 and is normally expressed in units of m^2 or

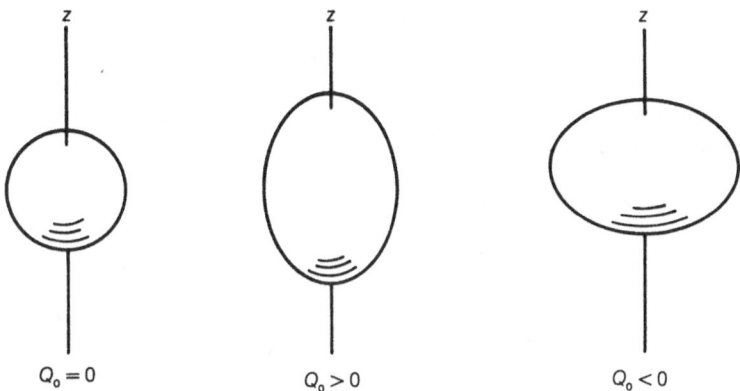

Fig. 2.5 Nuclear quadrupole shapes: spherical ($Q_0 = 0$), prolate ($Q_0 > 0$) and oblate ($Q_0 < 0$).

barns (where 1 barn $= 10^{-28}$ m^2). Clearly $\langle r^2 \rangle = \langle x^2 \rangle + \langle y^2 \rangle + \langle z^2 \rangle$ and also, if the nucleus is spherical, $\langle x^2 \rangle = \langle y^2 \rangle = \langle z^2 \rangle$ so that $\langle z^2 \rangle = \frac{1}{3} \langle r^2 \rangle$. Thus, for a spherical charge distribution $Q_0 = 0$ (Fig. 2.5).

Consider now a nucleus for which $\langle z^2 \rangle \neq \frac{1}{3} \langle r^2 \rangle$. There are two possibilities (Fig. 2.5):

$$\langle z^2 \rangle > \tfrac{1}{3} \langle r^2 \rangle \text{ (prolate ellipsoid) for which } Q_0 > 0$$
$$\langle z^2 \rangle < \tfrac{1}{3} \langle r^2 \rangle \text{ (oblate ellipsoid) for which } Q_0 < 0$$

and so positive and negative values of Q are to be expected.

The foregoing discussion is, however, essentially classical, and a proper quantum mechanical treatment requires some slight modification. In particular, ρ should be replaced by $\psi^* \psi$ where ψ is the nuclear wave function. This leads to the result for a nucleus of spin J that the actual measured nuclear electric quadrupole moment, Q, is now written (for $J > 0$)

$$Q = \frac{2J - 1}{2(J + 1)} Q_0 \tag{2.23}$$

For $J = 0$, Q vanishes identically because no symmetry axis is defined and it will be noted that Q also vanishes for $J = \frac{1}{2}$. For very large J, i.e. near the classical limit in which the spin is much larger than $\hbar$, the quantum unit of angular momentum, $Q \simeq Q_0$ as should be expected.

Many nuclear quadrupole moments have been measured using, for example, optical hyperfine structure and atomic beam techniques and values in the approximate range from $-1 \times 10^{-28}\,\mathrm{m}^2$ to $+8 \times 10^{-28}\,\mathrm{m}^2$ are found. The values are particularly high for the rare earth nuclei, but details of this variation, which bear very significantly on our understanding of nuclear structure, are reserved to later on (section 4.2.3).

3

The internucleon potential

Throughout the previous chapter it has been assumed that there is an essentially attractive force – the nuclear force – between any two nucleons, responsible for holding stable nuclei together. This force is of necessity powerful since it must hold the nucleons together in a very small volume as well as overcome the disruptive effect of the repulsive Coulomb force between protons. Normally discussion of the force, $F(r)$, is in terms of the related internucleon potential, $V(r)$, where r is the vector separation between two nucleons and where force and potential are related by the usual expression $F(r) = -\nabla V(r)$.

In this chapter we first investigate the implications of nuclear properties for the form of $V(r)$. Some information can be obtained from simple qualitative considerations, but generally it is necessary to carry out very detailed theoretical analysis of experimental data on the two-nucleon system. We shall do this only to a limited extent, but in sufficient depth to extract the main features of the potential.

3.1 GENERAL FORM OF THE INTERNUCLEON POTENTIAL

We have seen (section 2.2) that the nuclear binding energy per nucleon (B/A) is approximately constant for stable nuclei, having a value around 8 MeV. This means that the energy required to remove a nucleon from a nucleus is essentially independent of the number of nucleons which it contains. The situation is therefore quite different from the one which obtains in the case of gravitational or electrostatic forces for which $V \propto 1/r$. In the gravitational case, for example, the potential energy of a unit mass at the surface of a gravitating sphere (i.e. the binding energy of unit mass) is

$$V = G\frac{M}{R} \tag{3.1}$$

where G is the gravitational constant, M is the mass of the sphere and R is its radius. Since M is proportional to the volume of the sphere (i.e. $\propto R^3$) it follows that the larger the mass the greater is the binding energy ($\propto R^2$); it is easier for a rocket to escape from the moon than from the earth. This arises because the gravitational potential between two elements of matter dies away very slowly as r increases ($V \propto 1/r$) so that the unit mass at the surface of a gravitating sphere experiences the attraction of all elements of the sphere.

The approximate constancy of B/A for nuclei implies that the internucleon potential cannot be long range and that its dependence on r must be very different from $1/r$. The obvious conclusion is that it must be a short-range potential such that a nucleon in the nucleus only experiences the attraction of a few neighbouring nucleons (this is referred to as **saturation** of the nuclear force). The number of neighbours will be roughly the same whatever the size of the nucleus and hence the binding energy of a nucleon (B/A) will be essentially independent of the number of nucleons in the nucleus, as is observed. Since, as estimated in section 2.1, the average separation of nucleons in the nucleus is around 1.8 fm it is reasonable to suppose

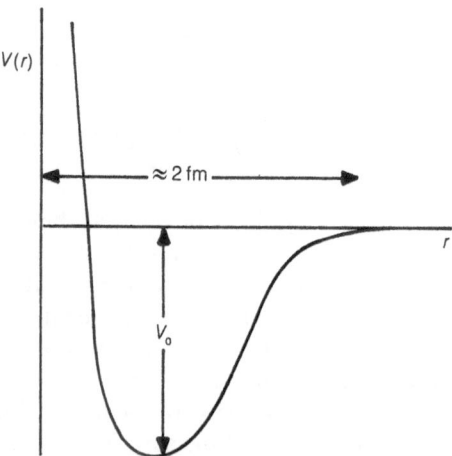

Fig. 3.1 Schematic shape of the internucleon potential.

that the range of the internucleon potential is of a similar order of magnitude (say 2 fm).

A further property of V follows from the fact that the volume of a nucleus is proportional to the mass number A (section 2.1). This proportionality implies that, in spite of the attractive character of the potential, nuclei do not collapse as A becomes larger – the nucleons keep their distance. V must therefore have a very short-range repulsive component which effectively keeps the nucleons apart when they approach each other very closely. The shape of the potential must therefore be of the form illustrated in Fig. 3.1.

As far as the strength of the potential is concerned, which is measured by the depth of the potential well (V_0) in Fig. 3.1, it must certainly be of the order of magnitude of the binding energy per nucleon. That is, it is expected to be measured in MeV, if not tens of MeV. As will be seen in the detailed calculations of the next sections the latter is the case.

3.2 THE DEUTERON

In order to determine more precise information about the inter-nucleon potential, it is necessary to consider the behaviour and properties of two-nucleon systems. The most detailed information has come from painstaking studies of the way in which one nucleon is scattered by another over a wide range of energies from the very low through to hundreds of MeV. The analysis of this scattering is extremely complicated and outline results only will be quoted as a gloss on the further information which can be obtained by considering the properties of the only (just!) bound system of two nucleons – the deuteron.

The deuteron (the nucleus of what is colloquially known as heavy hydrogen) is a bound state of a neutron and a proton and is denoted symbolically by $^2_1$H or, simply, ^{2}H. The values of its key static properties are as follows:

binding energy $B = 2.23\,\text{MeV}$
spin $J = 1$
magnetic moment $\mu = 0.86\,\mu_\text{N}$
electric quadrupole moment $Q = 2.82 \times 10^{-31}\,\text{m}^2$

In order to account for these properties theoretically it is clearly necessary to solve the relevant Schrödinger equation making some assumption about the form of the internucleon potential. In the

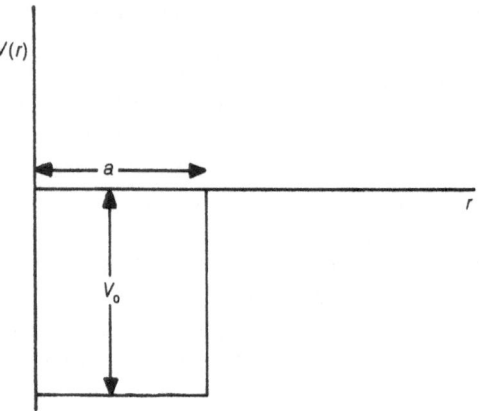

Fig. 3.2 Square well internucleon potential for the deuteron.

interests of simplicity we take the potential to be a square well of depth V_0 and width a (Fig. 3.2). This is clearly a crude approximation to a smooth curve of the shape conjectured in the previous section but it can nevertheless serve to give some indication of the depth and range of a more realistic potential. Denoting this square well potential by $V(r)$ the next problem is to solve the Schrödinger equation

$$\left[-\frac{\hbar^2}{2m}\nabla^2 + V(r) \right]u(r, \theta, \phi) = Eu(r, \theta, \phi) \tag{3.2}$$

where r, θ, ϕ are the internucleon spherical polar coordinates in the deuteron and $m\,(= m_{\text{p}}m_{\text{n}}/(m_{\text{p}} + m_{\text{n}}) \simeq m_{\text{p}}/2)$ is the reduced mass of the neutron–proton system. Since $V(r)$ is spherically symmetrical it follows in the usual way that the solutions of eq. (3.2) will have the general form

$$u_{nlm}(r, \theta, \phi) = R_{nl}(r)Y_{lm}(\theta, \phi) \tag{3.3}$$

where the quantum numbers n, l, m have their usual values:

$$n = 1, 2, 3, \ldots; \quad l = 0, 1, \ldots, n - 1; \quad m = l, l - 1, \ldots, -l$$

The corresponding energy eigenvalues depend on n and l and will be denoted by E_{nl}.

Note: it is now conventional to use the **radial** quantum number, n_{r}, in nuclear physics rather than the **principal** quantum number, n_{p}, of atomic physics. $n_{\text{r}} - 1$ specifies the number of nodes in the radial function and is related to n_p by $n_p = n_r + l$. n_r is simply denoted by n in this book.

3.2.1 The Deuteron Ground State Configuration

The deuteron binding energy is relatively small and experiment shows that it has, in fact, no excited states. We therefore only need to consider the lowest energy eigenstate of the Schrödinger equation which, for a square well potential, corresponds to the lowest possible values of the quantum numbers, i.e. $n = 1, l = 0, m = 0$. This is referred to as an S-state ($l = 0$) and since the neutron and proton both have spin $\frac{1}{2}$ there are two possibilities for the S-state configuration, namely $^3S_1$ (triplet) or $^1S_0$ (singlet) corresponding to total angular momentum quantum numbers 1 and 0 respectively. But the nuclear spin, J, is simply the total angular momentum quantum number and since, experimentally, we know that $J = 1$ it must be presumed that the deuteron is in the $^3S_1$ state.

This supposition is confirmed by considering the magnetic moment of the deuteron. In an S-state there is no contribution due to orbital motion of the charged proton and the magnetic moment must derive from the intrinsic magnetic moments of the component neutron and proton. As discussed in section 2.7 the magnetic moment is evaluated for the magnetic substate $M = J = 1$. In this state both neutron and proton must have $m_s = \frac{1}{2}$ (to give $M = m_s(n) + m_s(p) = 1$) – their intrinsic spins, and therefore their magnetic moments, are aligned. The deuteron magnetic moment, μ, is thus predicted to be simply equal to the sum of the neutron and proton magnetic moments, i.e.

$$\mu = \mu_n + \mu_p$$
$$= (-1.91 + 2.79)\,\mu_N$$
$$= 0.88\,\mu_N$$

to be compared with the experimental value $\mu = 0.86\,\mu_N$. There is obviously very good agreement.

3.2.2 The Deuteron Ground State Energy

Assuming that $n = 1, l = 0$ for the ground state we now have to solve the Schrödinger equation. Since $l = 0$ there is no angular dependence ($Y_{00} = $ constant) and so we only have to consider the radial equation satisfied by R_{10}. Writing

$$R_{10} = \frac{v}{r} \tag{3.4}$$

standard Schrödinger equation manipulation leads to the following
equation for v:

$$-\frac{\hbar^2}{2m}\frac{d^2v}{dr^2} + V(r)v = E_{10}v \qquad (3.5)$$

For a bound state E_{10} is, of course, negative and equal to $-B$ where B
is the binding energy.

In solving eq. (3.5) two regions of space have to be considered,
namely, $r < a$, where $V(r) = -V_0$, and $r > a$, where $V(r) = 0$. Taking
these two regions in turn, solutions emerge as follows.

For $r \leqslant a$, $V = -V_0$, and eq. (3.5) simplifies to

$$-\frac{\hbar^2}{2m}\frac{d^2v}{dr^2} = (V_0 - B)v \qquad (3.6)$$

and the general solution for this region of space is

$$v_< = A\sin\alpha r + C\cos\alpha r \qquad (3.7)$$

where

$$\alpha = \sqrt{\left[\frac{2m(V_0 - B)}{\hbar^2}\right]} \qquad (3.8)$$

and A and C are constants. The C term leads to a divergence in
$R_{10}(=v/r)$ when $r \rightarrow 0$ and we must therefore take $C = 0$.

For $r > a$, $V = 0$, and eq. (3.5) simplifies to

$$-\frac{\hbar^2}{2m}\frac{d^2v}{dr^2} = -Bv \qquad (3.9)$$

and the general solution for this region of space is therefore

$$v_< = De^{-\beta r} + Fe^{+\beta r} \qquad (3.10)$$

where

$$\beta = \sqrt{\left(\frac{2mB}{\hbar^2}\right)} \qquad (3.11)$$

and D and F are constants. Since the F term diverges as $r \rightarrow \infty$ it is
unacceptable as part of a wave function and we must therefore choose
$F = 0$.

The final step in this calculation is to impose the condition that v
and dv/dr are continuous at $r = a$.

Using the expressions for $v_<$ and $v_>$ given in eqs (3.7) and (3.10)
with $C = F = 0$ two equations result from this continuity

requirement:

$$A \sin \alpha a = De^{-\beta a}$$

and

$$\alpha A \cos \alpha a = -\beta De^{-\beta a}$$

Dividing the second by the first then gives

$$\cot \alpha a = -\frac{\beta}{\alpha}$$

or, using eqs (3.8) and (3.11) for α and β

$$\cot \alpha a = -\sqrt{\left(\frac{B}{V_0 - B}\right)} \tag{3.12}$$

There are two unknowns in this equation – a and V_0 – and a relation between them can be found using numerical methods. However, as a reasonable first approximation it can be assumed that, since B is relatively small compared with the average binding energy per nucleon in most nuclei, it can be neglected compared with V_0 so that eq. (3.12) reduces to

$$\cot \alpha a = 0 \tag{3.13}$$

where, now $\alpha \simeq (2mV_0/\hbar^2)^{\frac{1}{2}}$. Equation (3.13) implies that $\alpha a = \pi/2$, $3\pi/2$, $5\pi/2$, etc. But since we are dealing with the ground state, we must take $\alpha a = \pi/2$ since higher values would correspond to more oscillations in the wave function (eq. (3.7)) and, therefore, higher energy. Using the approximate expression for α above it therefore follows that

$$V_0 a^2 \simeq \frac{\pi^2 \hbar^2}{8m} = \frac{\pi^2 \hbar^2}{4m_{\mathrm{p}}} \simeq 10^{-28} \, \mathrm{MeV \, m^2} \tag{3.14}$$

3.3 DETAILED FORM OF THE INTERNUCLEON POTENTIAL

We have seen in the previous section that the wave function for the deuteron, assuming a square well internucleon potential, has the form (see eqs (3.7) and (3.10) with $C = F = 0$)

$$v_< = A \sin \alpha r \text{ for } r \leqslant a$$
$$v_> = De^{-\beta r} \text{ for } r > a$$

where α and β are given in eqs (3.8) and (3.11). In particular, using the experimental value for B, $\beta = 2.33 \times 10^{14} \, \mathrm{m}^{-1}$ or $\beta^{-1} =$

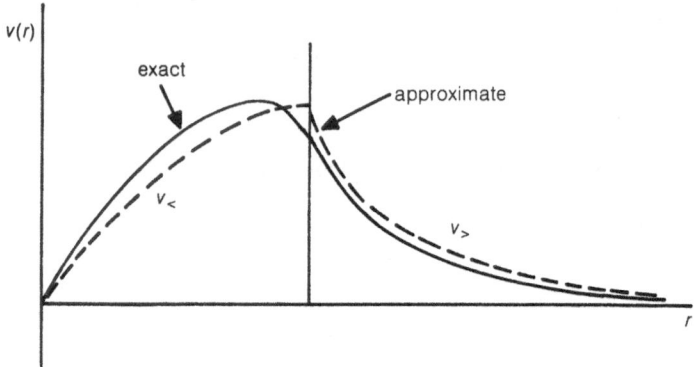

Fig. 3.3 The deuteron wave function in a square well potential.

4.30×10^{-15} m $= 4.30$ fm. This latter figure gives a scale for the distance over which the deuteron wave function dies away – it gives an indication of the 'size' of the deuteron and it is clearly much larger than the approximate range of the internucleon potential conjectured in section 3.1, namely $\simeq 2$ fm. This situation is shown in Fig. 3.3. The 'approximate' curve refers to the approximate solution to eq. (3.12) just discussed whilst the 'exact' curve refers to the exact solution to this equation and a has been taken to be 2 fm. It is clear that the neutron and proton in the deuteron spend much of their time outside the range of the internucleon potential.

Continuing with the assumption that $a \simeq 2$ fm we can now use eq. (3.14) to estimate V_0. Substituting for a gives at once $V_0 \simeq 25$ MeV in conformity with earlier expectations about the strength of the potential in section 3.1 and with the assumption in the foregoing approximate calculations that B is much smaller than V_0. Solving eq. (3.12) exactly using the same value for a leads to the even larger value $V_0 \simeq 35$ MeV.

A further conclusion can be drawn from the data considered so far, namely that the internucleon potential must be spin dependent. This follows because in the deuteron ground state the neutron and proton are in a triplet spin state ($J = 1$) and there are no excited states. If the potential was independent of the relative spin orientations of neutron and proton then a singlet deuteron state ($J = 0$) would exist with the same energy as the triplet state. This is not observed and the absence of such a state implies that the singlet potential is certainly weaker than the triplet potential.

Further information about this spin dependence can, as mentioned earlier, best be obtained by studying scattering processes. For the scattering of low energy (a few MeV) neutrons by protons only S-wave ($l = 0$) interactions need be considered since neutrons with $l > 0$ would be well outside the nuclear force range. Analysis of scattering data in terms of a square well potential then gives the following results (e.g. Brown and Jackson, 1976):

1. For triplet (t) interaction ($J = 1$),

$$V_{0t} \simeq 31.3 \, \text{MeV} \qquad a_t \simeq 2.21 \, \text{fm}$$

2. For singlet (s) interaction ($J = 0$),

$$V_{0s} \simeq 13.4 \, \text{MeV} \qquad a_s \simeq 2.65 \, \text{fm}$$

These values indicate clearly the difference in strength and range of the triplet and singlet potentials. Further calculations show that the singlet state of the deuteron is, in fact, very nearly bound but not quite; it misses being bound by around 0.1 MeV.

This difference between the triplet and singlet potentials means that any mathematical expression for the internucleon potential must also involve the spin operators s_1 and s_2 for the two nucleons. To conserve angular momentum and parity the potential must also be a scalar quantity, and the simplest possibility is

$$V = V_C + V_S = f_C(r) + f_S(r)s_1 \cdot s_2 \tag{3.15}$$

where f_C and f_S are two different functions of $r(= r_{12})$ representing the form and strength of the central (C) and spin-dependent (S) parts of the potential. We have so far taken these to be square wells but, in reality, they will be smooth curves along the lines indicated in Fig. 3.1. $s_1 \cdot s_2$ has the eigenvalues $\hbar^2/4$ and $-3\hbar^2/4$ for $J = 1$ and 0 respectively and therefore leads to different strengths for the singlet and triplet states (problem 3.2).

The internucleon potential is thus turning out to be a little complicated and further complications follow! For example, no account has been taken so far of the fact that the deuteron has a small positive electric quadrupole moment Q. As discussed in section 2.7, this implies that the deuteron does not have spherical symmetry but is a prolate ellipsoid. But, so far, it has been taken to be in a $^3S_1$ state in which the wave function has no angular dependence and is therefore spherically symmetrical with $Q = 0$. To account for the quadrupole moment some angular dependence has to be introduced and the wave function must have another component. There are three possibilities, i.e. $^3P_1$ ($S = 1$, $L = 1$, $J = 1$), $^1P_1$ ($S = 0$, $L = 1$, $J = 1$) and $^3D_1$ ($S = 1$, $L = 2$, $J = 1$), all of which have $J = 1$ as required. The first two can be discounted at once since P-states have odd parity (for a two-nucleon system parity $= (-1)^l$) whilst an S-state has even parity and, as discussed in section 2.6, nuclear states have a definite parity so that a mixture of parities is not allowed. Thus, only the $^3D_1$ state can be admixed and the deuteron wave function must have the form

$$\Psi = (1 - p)^{1/2}\,\Psi(^3S_1) + p^{1/2}\,\Psi(^3D_1) \qquad (3.16)$$

where p is the probability of the deuteron's being in the $^3D_1$ state. The measured quadrupole moment is accounted for if $p \simeq 5$–8%. This admixing of S- and D-states implies that there must be a component of the internucleon potential able to bring this about. The situation is illustrated in Fig. 3.4 where, since $Q > 0$, this component must lead to configuration (a) having lower energy than configuration (b). In qualitative terms it must be energetically more favourable for the nucleon spins to be oriented parallel rather than perpendicular to r. The additional contribution to the nuclear force responsible for this is

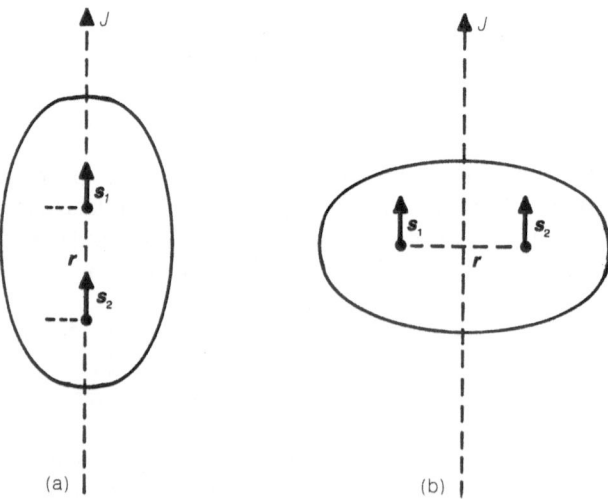

Fig. 3.4 In the deuteron the tensor force favours the prolate shape (configuration (a)) over the oblate shape (configuration (b)).

known as the **tensor force** and the corresponding potential has the form

$$V_T = f_T(r)\left[\frac{3(s_1 \cdot r)(s_2 \cdot r)}{r^2} - s_1 \cdot s_2\right] \qquad (3.17)$$

where f_T is a function giving the strength and radial dependence of the potential. This potential is similar in form to the potential between two magnetic dipoles but is physically quite unrelated. It can be seen that it will have different values according to the nucleon spins being parallel or perpendicular to r.

Extensive programmes of nucleon–nucleon (n, p) and (p, p) scattering (it is physically impossible to make a direct study of (n, n) scattering) at energies up into the GeV range together with detailed theoretical analyses have been carried out over many years and there are now available very detailed representations of the internucleon potential which can account for the experimental data. The potential turns out to be extremely complicated including not only terms of the type that we have been discussing but also many other contributions. For example, experiments using polarized nucleons (i.e. with their spins oriented in a particular direction) indicate the presence of a spin–orbit potential of the form

$$V_{LS} = f_{LS}(r)L \cdot (s_1 + s_2) \qquad (3.18)$$

where L is the relative orbital angular momentum operator for the two nucleons. Another interesting phenomenon in the scattering of high energy neutrons by protons and referred to as **exchange** is also observed. Suppose a neutron with energy 500 MeV collides with a proton at rest in the laboratory. Since the nuclear potential is now small compared with the neutron energy it would be expected that, although a little energy would be transferred to the proton, in general the neutron would continue along its path in the forward direction with most of its energy. In fact, it is found that the proton is just as likely to be scattered in this direction with a high energy. It is as though the neutron picked up the proton's charge, converting itself into a proton and leaving the original proton behind as a neutron! This phenomenon must also be accounted for in any representation of the internucleon potential. *In toto*, then, the potential has a very complicated structure:

$$V = V_C + V_S + V_T + V_{LS} + \ldots \qquad (3.19)$$

where ... indicates the presence of other terms. In spite of the complications various successful phenomenological representations have been produced (Brown and Jackson, 1976) which can account in detail for the observed interactions between two nucleons.

3.3.1 Charge Independence

A final feature of the internucleon potential to be noted is that it is essentially 'charge independent'. That is, it has the same form and strength whether it is two neutrons, two protons or a neutron and proton that are interacting provided that they are in the same state. Here it should be remembered that there are neutron–proton states not available to two neutrons or two protons because of restrictions imposed by the Pauli exclusion principle. For example, a neutron and a proton can be in a $^3S_1$ state, as in the deuteron ground state, but this is not allowed for two neutrons or two protons. This charge independence means that, as far as the internucleon potential is concerned, protons and neutrons have the same behaviour although they are, of course, radically different in their electromagnetic properties.

For this reason, the nucleons are allocated an **isospin** $\frac{1}{2}$ with the z components $+\frac{1}{2}$ and $-\frac{1}{2}$ corresponding to the proton and neutron respectively. Isospin is only mathematically related to ordinary spin and has nothing to do with physical angular momentum as such. It is simply a convenient way of dealing with charge independence, which can be embodied in the statement that the internucleon potential is independent of the z component of isospin; the potential must be a scalar in isospin space (section 7.3 contains a fuller discussion).

As far as nuclear structure is concerned, charge independence manifests itself as an approximate correspondence between the energy levels of nuclei which have the same number of nucleons, but different numbers of neutrons and protons. The correspondence is approximate because of Coulomb effects which, of course, are charge dependent. It shows itself, for example, in the spacing and spin–parities assignments of the levels in ^{17}O and ^{17}F shown in Fig. 2.4. These nuclei which both have $A = 17$ but with $Z = 8$, $N = 9$ and $Z = 9$, $N = 8$ respectively are referred to as **mirror nuclei**. In general, corresponding states are said to belong to an **isospin multiplet**.

3.4 MESON THEORY OF THE INTERNUCLEON POTENTIAL

The contrast between the simplicity of the Coulomb potential and the complexity of the internucleon potential is striking. Not only does the latter have a complicated dependence on the vector separation between nucleons as well as their spins but it is also characterized by a range ($\simeq 2$ fm) whilst the former ($V \propto 1/r$) has infinite range. The Coulomb potential is well understood classically in terms of the electromagnetic field between two charged particles. From Maxwell's equations we know that an electromagnetic field is transmitted as a wave motion with the speed of light, c. From a quantum mechanical perspective this transmission is described in terms of massless photons of momentum p and energy $E(= pc)$. If the wavelength and frequency of the equivalent wave motion are λ and v respectively then the following well-known equations hold:

$$c = v\lambda$$
$$E = hv \tag{3.20}$$
$$p = h/\lambda$$

Considering, for example, the scattering of one electron (charge e) by another, the process can be represented symbolically as in Fig. 3.5. In (a) the interaction between them is represented by a classical electromagnetic field and in (b) by the exchange of a photon. Calculations on either of these two bases lead to essentially the same result for the potential responsible for the scattering, namely,

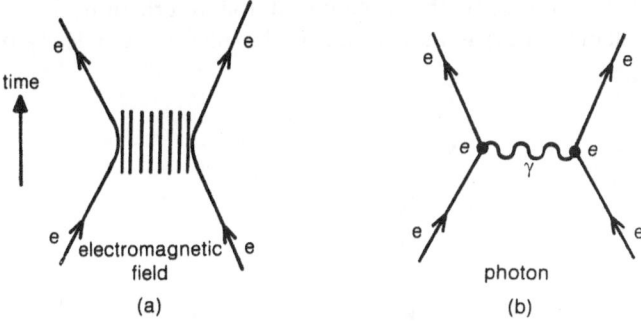

Fig. 3.5 Electron–electron scattering: (a) classical (field) description; (b) quantum (photon exchange) description.

$V = e^2/4\pi\varepsilon_0 r$. In the quantum mechanical calculation the infinite range of the potential is seen as due to the zero mass of the photon.

In 1935, a Japanese physicist – Hideki Yukawa – suggested that an interaction of finite range, as needed in nuclear physics, could be generated if particles (now referred to as **mesons**) of finite mass were exchanged between nucleons. Of course, if a nucleon is to emit a particle of mass m (say) then energy of at least mc^2 is needed to create it. This is no problem as long as this energy (denoted by ΔE) is repaid within a time Δt given by the uncertainty relation

$$\Delta E \, \Delta t \simeq \hbar$$

Since the energy has to be repaid, the meson is referred to as a 'virtual' particle. Substituting $\Delta E = mc^2$ then gives for this time

$$\Delta t \simeq \frac{\hbar}{mc^2}$$

During time Δt the furthest distance that the meson can travel, assuming a speed of the order c (the velocity of light), is $c \, \Delta t = \hbar/mc$. This means that the range (R) of the interaction between two nucleons due to meson exchange (Fig. 3.6) is therefore

$$R \simeq \frac{\hbar}{mc} \tag{3.21}$$

Thus, if $R \simeq r_0 \simeq 2\,\text{fm}$, we require $m \simeq 200\, m_e$, a value intermediate between the mass of an electron (m_e) and the mass of a nucleon $(\simeq 1840\, m_e)$ – hence the name 'meson'.

Continuing with this 'meson exchange' hypothesis it is also necessary to specify the strength of the interaction through an equivalent quantity to the charge e used to measure the strength of the

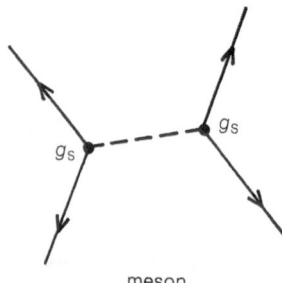

Fig. 3.6 Meson exchange between two nucleons.

electromagnetic interaction. This is usually denoted by g (Fig. 3.6) and (like e) is referred to as a 'coupling constant' since it measures the strength of the 'coupling' of the meson to the nucleon (just as e measures the strength of the coupling of a photon to a charged particle). Yukawa, in detailed calculations based on this hypothesis, then obtained an internucleon potential of the form

$$V(r) = \frac{g_S^2}{4\pi} \frac{e^{-r/R}}{r} \tag{3.22}$$

where R is now given precisely by $R = \hbar/mc$. Here it should be noted that the Yukawa potential reduces to the Coulomb form ($\propto 1/r$) in the limit $m \to 0$ ($R \to \infty$ so that $e^{-r/R} \to 1$).

A rough estimate of the size of g can be obtained by taking $R = 2\,\text{fm}$, evaluating the magnitude of $V(r)$ at $r = 2\,\text{fm}$ and equating this to the depth of the square well potential already estimated in section 3.3 – say, 30 MeV. A simple calculation gives $g_S^2/4\pi \simeq 10^{-26}\,\text{J m}$ or, putting this result in dimensionless form,

$$\alpha_S = \frac{g_S^2}{4\pi\hbar c} \simeq 1 \tag{3.23}$$

where α_S measures the strength of the interaction. This is to be compared with the corresponding dimensionless quantity measuring the strength of the electromagnetic interaction, namely, $\alpha = e^2/4\pi\varepsilon_0\hbar c \simeq 1/137$ – the fine structure constant. The former is two orders of magnitude larger than the latter and, for this reason, the internucleon potential is referred to as a 'strong interaction'. For future reference (Chapter 8) the strength of the strong interaction will be taken to be $\alpha_S \simeq 1$.

3.4.1 Pions

Of course, in 1935, the foregoing approach to understanding the internucleon potential was simply a hypothesis and its essential correctness was only confirmed in 1947 when particles having the required properties of the exchanged mesons were discovered by Lattes, Muirhead, Occhialini and Powell. They exposed photographic plates at high altitudes and found cosmic ray tracks of charged particles in them whose mass was of the required order of magnitude. The mass could be estimated from the nature of the tracks in the emulsion, which also indicated that the particles interacted strongly with the nuclei in the emulsion as required. The particle was named the π meson and is now normally referred to as the pion. Subsequent detailed work established the following facts about pions.

1. There are three types – π^+, π^0 and π^- – with charges $+e$, 0 and $-e$ respectively.
2. Their masses are as follows:

$$m(\pi^+) = m(\pi^-) = 273.1\, m_e = 139.57\, \text{MeV}/c^2$$
$$m(\pi^0) = 264.1\, m_e = 134.97\, \text{MeV}/c^2$$

3. They all have intrinsic spin 0 and intrinsic odd parity – they are therefore labelled as 0^- particles.
4. All three particles are highly unstable. The π^+ and π^- decay on average in a time of the order of 10^{-8} s, mainly emitting a muon (a 'heavy' electron) and a neutrino, whilst the π^0 decays electromagnetically in a time of the order of 10^{-16} s, emitting two photons (Table 7.1 and Chapter 9 provide details of these decay processes).

In terms of pion exchange there are several diagrams that can be drawn representing the interaction between two nucleons (Fig. 3.7).

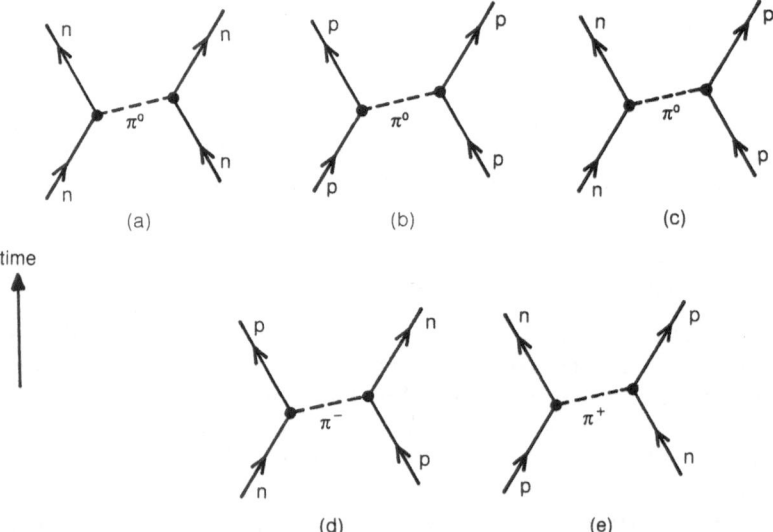

Fig. 3.7 Pion exchange between two nucleons. (a), (b) and (c) involve π^0 exchange whilst (d) and (e) involve $\pi^\pm$ exchange leading to exchange scattering.

The corresponding potentials can be calculated and are found to feature many of the general spatial properties required: short range, spin dependence, tensor term. In addition, the exchange character of the internucleon potential can easily be understood in terms of diagrams (d) and (e) in Fig. 3.7 in which charge is seen to be exchanged between neutron and proton via the π^- and π^+. Further, since pions like nucleons are virtually identical as far as their strong interactions are concerned (note the near equality of their masses), the resultant potential is charge independent as required. However, pion exchange only accounts for the behaviour of the potential at distances greater than about 1 fm. At shorter distances the exchange of heavier mesons (e.g. ρ meson ($m_\rho = 770\,\text{MeV}/c^2$) and ω meson ($m_\omega = 783\,\text{MeV}/c^2$) – Chapter 7) and two-pion exchange have to be invoked. In these terms a general understanding of the internucleon potential has been achieved.

4

Models of nuclear structure

We have seen in the last chapter that the internucleon potential is extremely complicated. Any full theory of nuclear structure therefore has to contend not only with a quantum mechanical many-body problem involving tens or, for heavier nuclei, hundreds of nucleons but also with their interaction with each other through this complicated potential. An exact treatment is clearly not possible and the approach over the last half-century has been to devise increasingly sophisticated conceptual models of the nucleus which are simple enough to enable calculations to be made, but near enough to the physical situation to enable reasonably detailed understanding of nuclear behaviour to be achieved. Different models focus on different aspects of nuclear behaviour (ground state energy, excited state energies, electromagnetic properties, nuclear reactions at different energies, etc.) and on different groups of nuclei. In this chapter we shall consider the basic models which have facilitated understanding of nuclear energy levels (their energies, spins and parities) and their electromagnetic properties (magnetic dipole and electric quadrupole moments).

4.1 THE LIQUID DROP MODEL

This model enables a general overview of the masses, binding energies and stabilities of the full spectrum of nuclei to be obtained. It is based on the simple idea that a nucleus behaves in some respects like a drop of liquid. For example, intermolecular forces are relatively short ranged so that the amount of energy required to evaporate a given mass of liquid from a drop (latent heat of vaporization) is independent of the size of the drop just as the binding energy of a nucleon is roughly independent of A (section 2.2). In terms of this physical picture as a basis it is then straightforward to introduce different contributions to the total mass of a nucleus. The forms of these terms are determined by physical considerations and they are expressed in terms of parameters whose values are determined by comparison with experimental data on nuclear masses. For a nucleus with atomic number Z and mass number A there are the following contributions to the mass of the corresponding neutral atom.

1. Mass of constituent nucleons: there are Z protons and $A - Z$ neutrons so that

$$M_N = Zm_p + (A - Z)m_n$$

2. Volume binding energy: in the volume of the nucleus each nucleon experiences the attraction of a few neighbouring nucleons only (because of the short-range character of the internucleon potential). Because of the attractive nature of the potential this is a negative contribution to the mass which is roughly the same for each nucleon and so the total contribution is simply proportional to A giving

$$M_V = -a_V A$$

where a_V is a parameter.
3. Surface effect: nucleons near the surface of the nucleus are surrounded by fewer nucleons and will therefore experience less attractive potential energy than those inside the nucleus. To compensate for this a positive term must be included proportional to the number of nucleons in the nuclear surface, i.e. proportional to the surface area ($\propto R^2 \propto A^{2/3}$), giving

$$M_S = +a_S A^{2/3}$$

where a_S is a parameter.
4. Coulomb energy: the nucleus has a total charge Ze essentially confined to a sphere of radius R. The resultant (positive) potential energy given by classical electrostatic theory for a uniform charge distribution is $(3/5)(Ze)^2/4\pi\varepsilon_0 R$. Extracting the key dependence on Z and R ($\propto A^{1/3}$ – eq. (2.2)) we can therefore use

$$M_C = a_C \frac{Z^2}{A^{1/3}}$$

where a_C is a parameter.

5. Asymmetry energy: it can be seen from Fig. 2.3 that stable light nuclei are characterized by $N \simeq Z$ (i.e. $A \simeq 2Z$), the trend to $N > Z$ for heavier nuclei being attributable to the increasing importance of Coulomb repulsion between the protons. This latter effect has just been taken into account and therefore a term reflecting the former effect – referred to as asymmetry energy – should be included. It should have the property that if A deviates from $2Z$ then there is a positive contribution to the energy. A simple way of doing this is through the expression

$$M_{As} = a_{As} \frac{(A - 2Z)^2}{A}$$

where a_{As} is a parameter.

6. Odd – even effect: it has been remarked in section 2.3 (Table 2.2) that the most stable nuclei have Z even and N even with, therefore, A even (even–even nuclei), the next most stable being odd–even or even–odd, with A odd, and the least stable being odd–odd, with A even. The precise form of a term reflecting this behaviour is a matter of choice, but the following (originally suggested by Fermi) is frequently used:

$$\begin{aligned}
\delta(A, Z) &= -a_{0-E} A^{-3/4} && \text{for } A \text{ even, } Z \text{ even} \\
&= 0 && \text{for } A \text{ odd} \\
&= +a_{0-E} A^{-3/4} && \text{for } A \text{ even, } Z \text{ odd}
\end{aligned}$$

where a_{0-E} is a parameter.

Gathering these different terms together and including the mass of the Z atomic electrons then gives the following expression (known as the Weizsäcker semi-empirical mass formula) for the mass of an atom:

$$M(A, Z) = Zm_p + (A - Z)m_n + Zm_e - a_V A + a_S A^{2/3}$$
$$+ a_c \frac{Z^2}{A^{1/3}} + a_{As} \frac{(A - 2Z)^2}{A} + \delta(A, Z) \qquad (4.1)$$

Fits to the experimental data have been made by various individuals and the following is a typical set of values for the various parameters (Green, 1954) expressed in terms of atomic mass units (u) and MeV/c^2:

	u	MeV/c^2
a_V	16.92×10^{-3}	15.76
a_S	19.12×10^{-3}	17.81
a_C	0.76×10^{-3}	0.71
a_{As}	25.45×10^{-3}	23.70
a_{0-E}	36.50×10^{-3}	34.0

The binding energy (B) of a nucleus is simply the energy difference between its actual mass and the mass of its constituent particles (eq. (2.5)) so that, from eq. (4.1), we have

$$B = [Zm_p + (A - Z)m_n + Zm_e - M(A, Z)]c^2$$

$$\simeq [a_V A - a_S A^{2/3} - a_C \frac{Z^2}{A^{1/3}}] c^2 \tag{4.2}$$

or,

$$\frac{B}{A} \simeq \left(a_V - a_S A^{-1/3} - a_C \frac{Z^2}{A^{4/3}} \right) c^2 \tag{4.3}$$

where the asymmetry and odd–even terms, which are fairly small, have been neglected. These three main contributions to B/A are plotted as a function of A (using appropriate values for Z) in Fig. 4.1 together with their total contribution. The latter can be seen to reflect the empirical properties of B/A shown in Fig. 2.2.

Similarly, an expression can be found for the value of Z, for a given A, which corresponds to the most stable nucleus. This is obtained by seeking the value of Z which gives a minimum in the expression given in eq. (4.1) for $M(A, Z)$. Taking $\delta(A, Z) = 0$ for simplicity, and putting $\partial M/\partial Z = 0$ gives, after substituting the numerical values for the various parameters,

$$Z = \frac{A}{1.97 + 0.015A^{2/3}} \tag{4.4}$$

This expression reflects the experimental observation that the valley of stability has $Z \simeq A/2 \simeq N$ for light nuclei with the proportion of protons decreasing as A increases (Fig. 2.3).

It should finally be stressed with this model that, although it represents very effectively the general trend of binding energies and stability considerations, it cannot be used to explore in any detail the

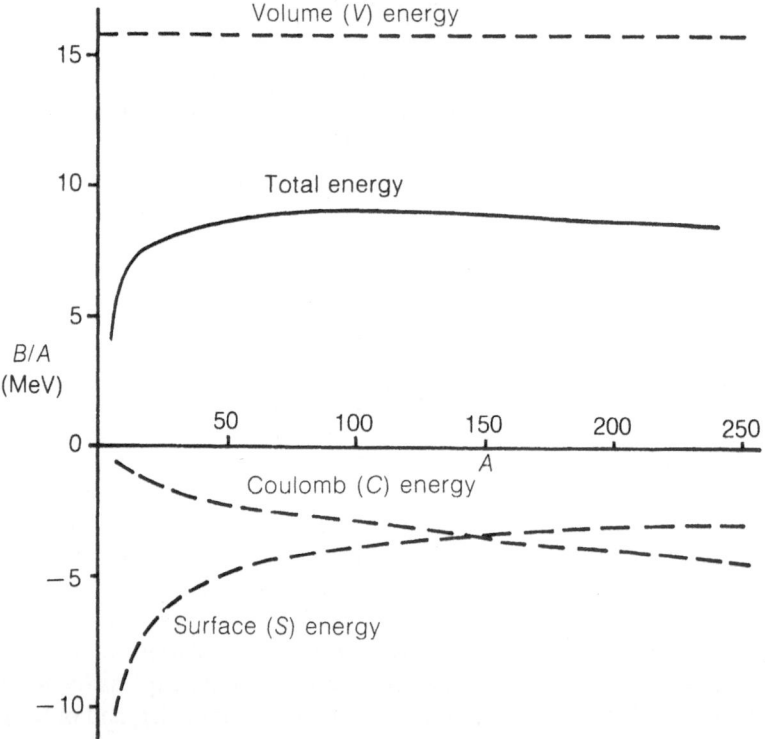

Fig. 4.1 Contributions of volume, surface and Coulomb energies to nuclear binding energy.

properties of an individual nucleus such as its spin, parity, electro-magnetic moments and energy levels. More sophisticated models have to be devised for this purpose.

4.2 THE NUCLEAR SHELL MODEL

The model which has acted as a basis for much of our understanding of nuclear structure is the shell model which takes into account the behaviour of individual nucleons in the nucleus. A rough approximation to this behaviour can be arrived at using the following considerations. Focusing attention on a single neutron as it moves through the body of the nucleus there is, on average, no net force on it due to other nucleons since they surround it fairly uniformly. It simply experiences a roughly constant negative (attractive) potential energy due to them. However, as it moves towards the nuclear surface, there will be an increasing attractive inward force leading to a decrease in its potential energy. This eventually reduces to zero as the neutron moves away from the nucleus and falls outside the range of the

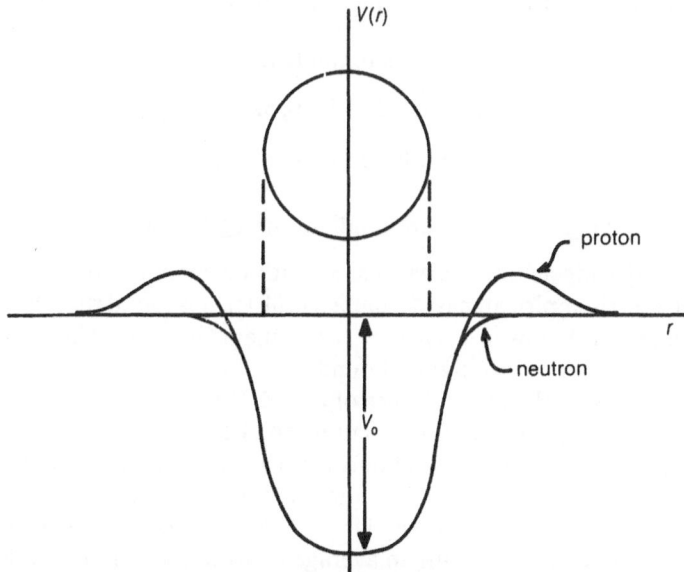

Fig. 4.2 Average potential energy experienced by a neutron or proton in a nucleus.

nuclear force. Exactly the same argument applies to a proton except that, as it moves away from the nucleus, it experiences a net electrostatic repulsion and therefore a positive contribution to the potential energy which only dies away slowly. This behaviour of the potential energy curve $V(r)$ is illustrated quantitatively in Fig. 4.2.

In such a potential well there will be a series of nucleon energy levels just as electron energy levels occur in an atom where, of course, the potential well is due to the positive electric charge carried by the nucleus. Again, as with the electrons in an atom, neutrons and protons in the nucleus can be accommodated in the energy levels according to the restrictions imposed by the Pauli exclusion principle which only allow a certain number of neutrons or protons in each level. It might therefore be expected that just as certain atoms with completely filled levels (closed shells) are particularly stable a similar phenomenon should be observed in nuclei. This is indeed the case.

4.2.1 The Magic Numbers

It is observed experimentally that nuclei having the following values of Z and N – known as the 'magic numbers' – have various distinctive qualities all indicating high stability:

Magic numbers

$$Z = 2, 8, 20, 28, 50, 82$$

$$N = 2, 8, 20, 28, 50, 82, 126$$

Some of the more important qualities are as follows.

1. Comparing actual nuclear masses with the smooth predictions of the semi-empirical mass formula (eq. (4.1)), it is found that they are significantly lower when Z or N is a magic number. These nuclei are therefore more tightly bound.
2. Nuclei with the above values of Z and N have significantly more stable **isotopes** (same Z, different N) and **isotones** (same N, different Z) respectively than neighbouring nuclei. For example, for $Z = 50$ there are ten stable isotopes compared with an average of around four for nearby nuclei. Similarly for $N = 20$ there are five stable isotones compared with an average of around two in that region.
3. The 'doubly magic' nucleus, ^{4}He, with $Z = N = 2$ is a particularly stable nucleus (Fig. 2.2) as are ^{16}O with $Z = N = 8$ and ^{208}Pb (the heaviest stable nucleus) with $Z = 82$ and $N = 126$.

4. Helium together with nuclei having $N = 50$, 82 and 126 are particularly abundant in the universe.
5. The first excited states of nuclei with a magic number of neutrons or protons are significantly higher in energy than for neighbouring nuclei – the nuclei are harder to disrupt.
6. Nuclear electric quadrupole moments, Q (section 2.7), are periodic as a function of Z or N, going through zero, i.e. a compact spherical shape, at the magic numbers.

The foregoing all suggest, by analogy with inert gases in the atomic situation, that at the magic numbers shells of nucleons have been just filled.

In order to obtain more specific results, some assumption must be made about the shape and strength of the potential well, $V(r)$, in Fig. 4.2. Estimates of its depth and spatial extent have been made by studying, for example, the way in which low energy neutrons interact with nuclei and these indicate that, roughly,

$$V_0 \simeq 50\,\text{MeV} \quad \text{and} \quad R \simeq 1.3A^{1/3}\,\text{fm}$$

Here it will be noted that R is somewhat larger than the radius of the density distribution given in eq. (2.2), reflecting the finite range of the nuclear force. In point of fact the depth and shape of the potential well are not critical for understanding level orderings and frequently an oscillator potential $(V \propto r^2)$ or an infinite square well have been used for illustrative purposes.

Staying with a diffuse potential well of the kind illustrated in Fig. 4.2 it is then a straightforward, if tedious, matter to solve the Schrödinger equation for a nucleon moving in such a potential. The eigenfunctions of this equation depend, as usual, on the three quantum numbers n, l and m_l whilst the eigenvalues, E, depend on n and l. The ordering of levels obtained is shown in Fig. 4.3 (s signifies $l = 0$; p, $l = 1$; etc.) and it will be noted that it is significantly different from the ordering in an atomic Coulomb potential. The levels bracketed on the left would be degenerate in an oscillator potential and the groups would be equally spaced.

The Pauli exclusion principle allows a level of given l to accommodate $2(2l + 1)$ identical nucleons ($(2l + 1)$ values of m_l with nucleon spin 'up' or 'down'). It is then a simple matter to show that although the magic numbers 2, 8 and 20 occur (after the filling of the 1s, 1p and 2s levels respectively) the remaining magic numbers cannot be generated (e.g. after filling the 1f and 2p levels the numbers 34 and 40 are obtained).

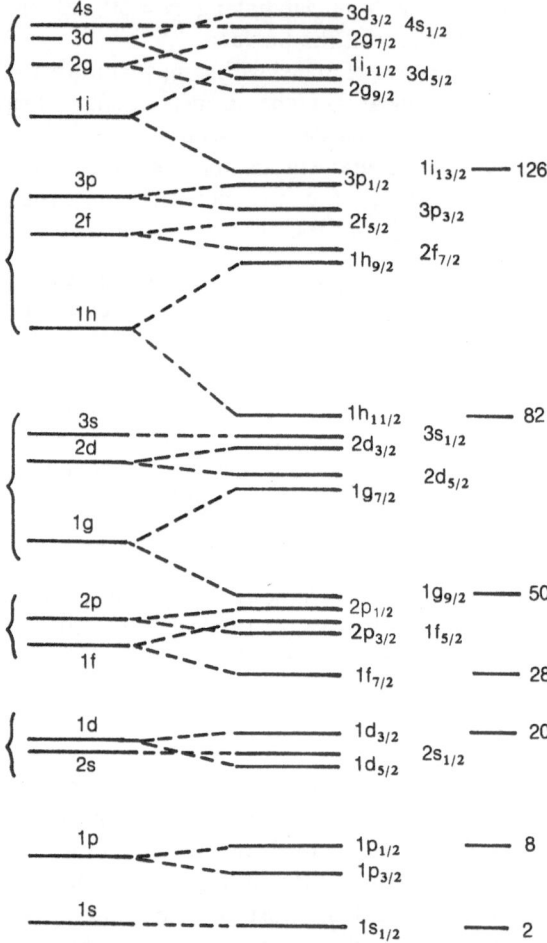

Fig. 4.3 Nuclear energy levels in a diffuse potential well. At the left are the levels obtained without spin–orbit coupling and, to the right, the levels obtained with spin–orbit coupling.

In 1948 a way of accounting for the magic numbers, which is now well established, was proposed independently by Mayer and Jensen. They suggested that, in addition to the static potential $V(r)$ experienced by a nucleon, there is also a strong spin–orbit potential of the form

$$V_{so}(r) = f(r)\mathbf{L}\cdot\mathbf{s} \tag{4.5}$$

where L and s are the nucleon orbital and spin angular momentum operators respectively. $f(r)$ is usually taken to be proportional to $(1/r)\,dV/dr$ where $V(r)$ is the potential well in which the nucleons move. This is similar to the spin–orbit term in atomic physics. However, nuclear spin–orbit coupling is nothing to do with electromagnetic effects and derives from an average of the complex spin-spatial dependence of the internucleon potential, in particular the two-body spin–orbit potential (eq. (3.18)). Its effect is to split each of the energy levels E_{nl} into two characterized by the total angular momentum quantum number $j = l \pm \frac{1}{2}$ (z component m). As in the atomic case the splitting, ΔE, is proportional to $2l + 1$ and, on average, experiment indicates (Bohr and Mottelson, 1969) that ΔE is approximately given by

$$\Delta E = E_{j=l-\frac{1}{2}} - E_{j=l+\frac{1}{2}} \simeq 10(2l + 1)A^{-2/3}\,\text{MeV} \qquad (4.6)$$

Clearly the splitting is much larger than in the atomic case and is also of the opposite sign with the $j = l - \frac{1}{2}$ level lying highest.

Returning now to the level diagram in Fig. 4.3, the effect of the splitting leads to a new series of energy levels (labelled nl_j) each of which can accommodate $2j + 1$ nucleons. Because the splitting increases with l it has a profound effect on the sequence of levels for $l = 3$ (f), 4 (g), 5 (h) and 6 (i), as shown in the diagram. The $1f_{7/2}$, $1g_{9/2}$, $1h_{11/2}$ and $1i_{13/2}$ levels join lower energy groups of levels so that filled shells now arise when the number of identical nucleons equals 28, 50, 82 and 126, the remaining magic numbers. In general there is a large energy gap between the level at the top of a filled shell and the next series of levels – hence the increased stability.

The task now is to see to what extent such a model of the nucleus can account for other experimental data.

4.2.2 Nuclear Spins and Parities. The Pairing Force

In terms of the shell model as formulated so far it is possible to make predictions about the spins and parities of the ground states of certain nuclei, namely those having completely filled shells or subshells of neutrons and protons and those having one nucleon more or less than the foregoing. The essential point is that a level with total angular momentum quantum number j is allowed by the Pauli exclusion principle to accommodate just $2j + 1$ like nucleons having z components of angular momentum equal to $m = j, j - 1, \ldots, -j$. The total value of the z component (Σm) is clearly 0, implying that the total angular momentum of these nucleons must be 0. Thus there is no contribution to the spin of a nucleus from closed subshells.

So for nuclei in which all subshells are full it follows at once that the nuclear spin is $J = 0$. Since $j(= l \pm \frac{1}{2})$ is half-integer, $2j + 1$ is necessarily even and such nuclei are even–even. The prediction for J is therefore in agreement with the general experimental observation (section 2.5) for even–even nuclei. Further, since the parities of all nucleons in a closed subshell are the same $((-1)^l)$ and there is an even number of them, it follows that the total parity of these nucleons must be $P = +1$ and hence the total parity of a nucleus having all its subshells closed must be even. For such nuclei, therefore, $J^P = 0^+$. Most nuclei in this category are **doubly magic**, for example ^4_2He, $^{16}_8\text{O}$, $^{208}_{82}\text{Pb}$, but there are others, for example $^{12}_6\text{C}$ ($1\text{s}_{1/2}$ and $1\text{p}_{3/2}$ subshells closed).

Consider now nuclei such as those just discussed ($J = 0^+$) but with an additional neutron or proton. Clearly the total angular momentum of such a nucleus and its parity must be simply those of the additional nucleon. For example $^{17}_8\text{O}$ has an additional neutron in the $1\text{d}_{5/2}$ state (Fig. 4.3) for which $j = 5/2$, $l = 2$ and $P = (-1)^l = +1$, leading to the prediction $J^P = \frac{5}{2}^+$ in agreement with experiment. For $^{41}_{20}\text{Ca}$ the odd neutron is in the $1\text{f}_{7/2}$ state and $J^P = \frac{7}{2}^-$, again in agreement with experiment. Similar results follow in the case of a nucleus with closed subshells less a nucleon. A subshell with a 'hole' in it has the same angular momentum as the missing particle. This follows since this angular momentum combined with that of the missing particle must give zero (we again have a filled subshell) and this can only be achieved if the two angular momenta are identical. Similarly the parity of a subshell with a 'hole' is the same as that of the missing particle. Thus for the nucleus $^{15}_7\text{N}$, which has a 'hole' in the $1\text{p}_{1/2}$ subshell, $J^P = \frac{1}{2}^-$. $^{207}_{82}\text{Pb}$ has a hole in the $3\text{p}_{1/2}$ sub-shell, giving again $J^P = \frac{1}{2}^-$, both predictions being in agreement with experiment.

For other nuclei a further consideration has to be introduced. Clearly, as formulated so far, the shell model is a crude approximation to reality. The sum total of interactions between nucleons has been simply approximated by the shell model potential $V(r)$ together with a single-particle spin–orbit potential. There must be a residual interaction between nucleons representing the difference between 'approximation' and 'reality'. This is normally referred to as an **effective** interaction and it should be included in some way. The nature of its main effect can be argued very simply.

Since the internucleon potential is basically attractive (apart from the repulsive core) it can be assumed that the effective interaction is also attractive. This means that the lowest energy state of a nucleus is one in which pairs of nucleons are as close together as possible so as to experience maximum attraction and therefore to contribute maximum negative potential energy. But, for like nucleons, the Pauli exclusion principle ensures that two nucleons in identical or similar states are well separated. It can therefore be argued that lowest energy in a given j subshell is achieved when the nucleons are paired together in completely dissimilar states, i.e. j, m with j, $-m$. This means that the angular momentum of each nucleon opposes that of the other so that the total angular momentum of the pair is zero. This pairing is of great importance in understanding nuclear structure and its influence has already been seen through the odd–even effect included in the semi-empirical mass formula (section 4.1) – even–even nuclei, having more pairs, are far more stable than odd–odd nuclei.

Accepting the pairing effect it then follows immediately that all even–even nuclei have $J^P = 0^+$, as is found experimentally, since all nucleons are paired off with each pair having zero angular momentum and, of course, even parity. Further, since an odd-A nucleus is simply an even–even nucleus together with an additional neutron or proton, it follows that the spin and parity of an odd-A nucleus should be simply equal to the angular momentum and parity of this single additional nucleon. In very many cases this prediction of the single-particle shell model agrees with experiment. However, there are exceptions of two types.

First, there are nuclei such as $^{107}_{47}\text{Ag}$ which is expected to have seven protons in the $1g_{9/2}$ state so that $J^P = \frac{9}{2}^+$ whereas experiment gives $\frac{1}{2}^-$. The reason for this and many similar anomalies is that, as can be shown by detailed calculations, the pairing energy increases with the value of j. It can thus be energetically more favourable to have eight (rather than seven) protons in the $1g_{9/2}$ state (an additional high j pair) together with one proton (rather than two) in the adjacent lower lying $2p_{1/2}$ state. The odd proton is now in the $2p_{1/2}$ state, giving the observed value for the spin and parity. Similar situations arise during the filling of the $1h_{11/2}$ and the $1i_{13/2}$ shells and this phenomenon accounts for the fact that the expected high ground state spins are not observed. They do arise, however, as low excited states and are referred to as 'isomeric' states (section 6.4.3).

Second, there are a few nuclei whose spins can simply not be accounted for in terms of a single odd nucleon. $^{23}_{11}\text{Na}$ is such a nucleus having three protons in the $1d_{5/2}$ state. With two paired off, J^P should be $\frac{5}{2}^+$ whereas it is observed to be $\frac{3}{2}^+$. All three protons must therefore contribute to the nuclear spin and more sophisticated versions of the shell model (section 4.4) are needed to account for this.

In the case of odd–odd nuclei, in which there is an odd neutron and an odd proton, recourse has again to be made to more detailed considerations and the model as formulated so far can give no definite prediction for the nuclear spin. But it is satisfying that such a simple model as the one being considered is able to account readily for the ground state spins and parities of most nuclei.

4.2.3 Electromagnetic Moments of Odd *A* Nuclei

In section 2.7 (eq. (2.20)) expressions for the magnetic moment operator and the value of the magnetic moment for a nucleus of spin *J*, in terms of the nuclear *g*-factor, have been deduced, namely:

$$\boldsymbol{\mu}_J = g_J \frac{e}{2m_p} \boldsymbol{J} \quad \text{and} \quad \mu_J = g_J \mu_N J \tag{4.7}$$

This means that even–even nuclei, having $J = 0$ in the ground state, will have no magnetic moment. But odd-*A* nuclei, all of which have half-integer spin, are expected to have a magnetic moment and its value, using the single-particle shell model, can be deduced very easily.

On this model, the magnetic moment is simply that of the odd nucleon in its appropriate shell model state characterized by the two quantum numbers *l* and *j*. There are contributions to the magnetic moment from the spin magnetic moment of the nucleon and also, in the case of protons, from the orbital motion. For a single nucleon the magnetic moment operator can, therefore, also be written:

$$\boldsymbol{\mu}_J = \frac{e}{2m_p}(g_L \boldsymbol{L} + g_S \boldsymbol{s}) \tag{4.8}$$

where the orbital *g*-factor, g_L, has the values $+1$ (proton) and 0 (neutron) and the spin *g*-factor, g_S, has the values $g_p = +5.5856$ (proton) and $g_n = -3.8262$ (neutron) as discussed in section 2.7. Equating the two expressions for $\boldsymbol{\mu}_J$ given in eqs (4.7) and (4.8) gives

$$g_J \boldsymbol{J} = g_L \boldsymbol{L} + g_S \boldsymbol{s}$$

or, on scalar multiplying through by *J*,

$$g_J \boldsymbol{J}^2 = g_L \boldsymbol{L} \cdot \boldsymbol{J} + g_S \boldsymbol{s} \cdot \boldsymbol{J} \tag{4.9}$$

But on the single-particle shell model the nuclear spin is simply the total angular momentum of the odd nucleon, so that

$$J = L + s$$

or

$$s = J - L \qquad (4.10a)$$

or

$$L = J - s \qquad (4.10b)$$

Squaring eqs (4.10a, b) gives expressions for $J \cdot L$ and $J \cdot s$ respectively in terms of J^2, L^2 and s^2 which can be substituted into eq. (4.9), giving

$$g_J J^2 = \tfrac{1}{2}[g_L(J^2 + L^2 - s^2) + g_S(J^2 + s^2 - L^2)] \qquad (4.11)$$

It is then a simple matter to replace each of the squared angular momentum operators by its eigenvalue (J^2 by $J(J+1)\hbar^2$, L^2 by $l(l+1)\hbar^2$ and s^2 by $\tfrac{1}{2}(\tfrac{1}{2}+1)\hbar^2$) and to obtain after some rearrangement

$$\mu_J = g_J J = \frac{1}{2}J\left[g_L + g_S + (g_L - g_S)\frac{l(l+1) - 3/4}{J(J+1)} \right] \qquad (4.12)$$

Using this equation and the values for g-factors already given, values for the magnetic moments of odd proton and odd neutron nuclei can easily be evaluated for the two cases $J = j = l \pm \tfrac{1}{2}$ (see problem 4.4). The results of such calculations are usually presented on what are known as Schmidt diagrams and these are shown in Fig. 4.4. The solid lines join the shell model predictions for the magnetic moments and are seen almost always to set bounds for the measured magnetic moments, a selection of which are included in the diagram. But it is clear that, although the single-particle shell model agrees with the general trend of the magnetic moments, in most cases it does not agree closely with experiment. The reason for this, as might be expected, is because of the extreme simplicity of the model, and it turns out that the magnetic moment predictions are actually very sensitive to small changes in the wave function due to the break up of pairs of nucleons. More sophisticated versions of the shell model (section 4.4) give much better agreement. But in some cases (e.g. for ^{3}H and ^{3}He) where the wave function is known with considerable accuracy, appeal has to be made to contributions to the magnetic moment from what are known as meson 'exchange' currents before good agreement between theory and experiment can be achieved. These currents arise, for example, from the motion of charged pions between nucleons and

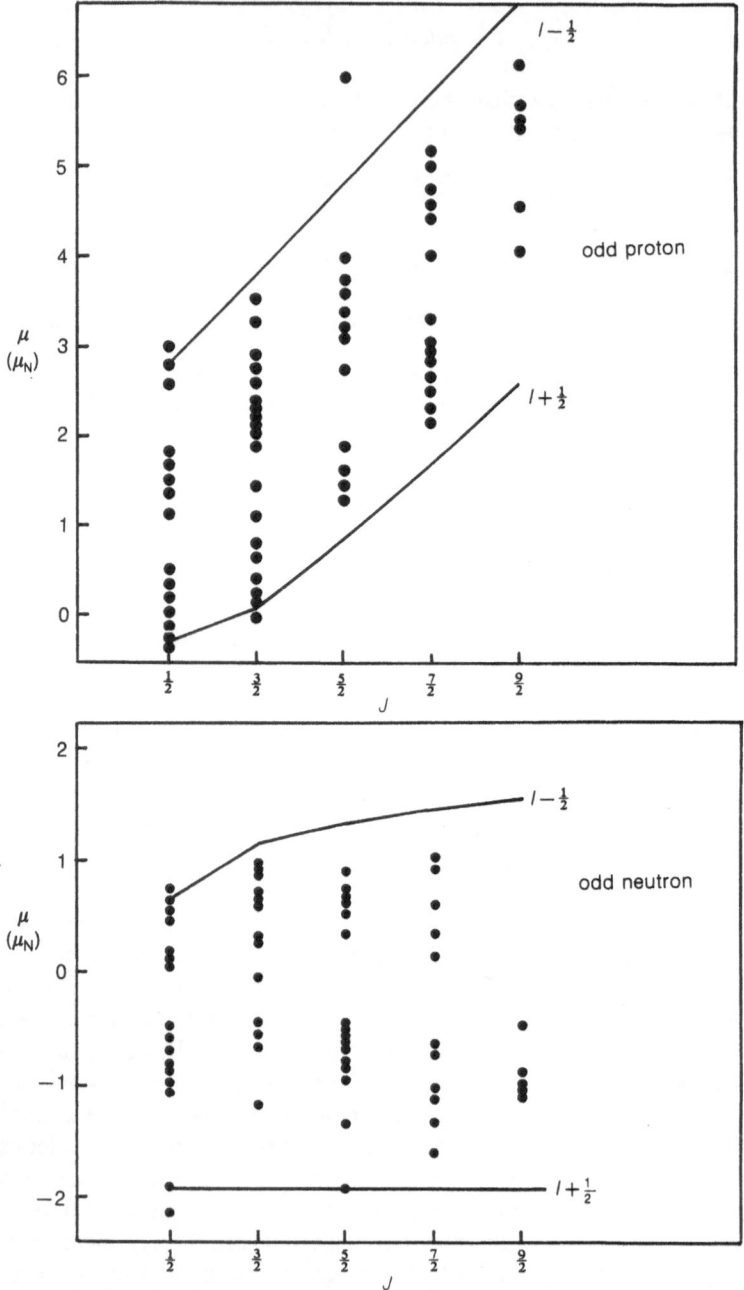

Fig. 4.4 Shell model predictions for the magnetic moments of odd-A nuclei (Schmidt lines) together with some experimental values.

which are responsible for part of the internucleon potential (section 3.4).

For the few stable odd–odd nuclei the situation is even more complicated, but some understanding of their magnetic moments can be achieved by treating the magnetic moment as the vector sum of the magnetic moments of each odd nucleon.

Turning now to electric quadrupole moments, Q, we have seen in section 2.7 that only nuclei with $J > \frac{1}{2}$ have non-zero values for Q and we shall therefore restrict consideration to odd-A nuclei. In the case of an odd proton nucleus, using the single-particle shell model, Q will be due simply to the odd proton and, ignoring the spin-dependent term, will have the order of magnitude (eqs (2.21) and (2.22))

$$Q_0 = (3\langle z^2 \rangle - \langle r^2 \rangle)$$

where Z in the expression for Q_0 has been put equal to 1 (single proton) and the averages are taken over the proton orbit. The averages will vary from around $\simeq 1 \times 10^{-30}\,\mathrm{m^2}$ for the lightest nuclei through to $\simeq 6 \times 10^{-29}\,\mathrm{m^2}$. In the case of odd neutron nuclei Q is predicted, apart from a very small recoil effect, to be zero since the neutron carries no electrical charge.

Experimentally a very different situation is found. First, there is no great difference between the values of Q for odd proton and odd neutron nuclei and, second, many nuclei are found to have values of Q up to an order of magnitude larger than the single-particle values. This happens in the approximate regions $150 < A < 190$ and $A > 225$, far away from closed shells. Such radical disagreements can only be explained if, in these regions, many nucleons contribute to the value of Q and we have here an indication of the phenomenon of 'collective' motion of nucleons in nuclei. In section 4.3 we shall see how such behaviour can be accommodated within the framework of another nuclear model – the collective model.

4.2.4 Excited States

Discussion so far has been about the ground states of nuclei and the extent to which the simple shell model can account for their properties. In the case of odd-A nuclei the natural expectation for low excited states is that they result simply from the excitation of the odd nucleon into a higher shell model state. For example the ground state of $^{17}\mathrm{O}$ has an odd neutron in the $1\mathrm{d}_{5/2}$ state and immediately above this lie the $2\mathrm{s}_{1/2}$ and $1\mathrm{d}_{3/2}$ states which might, therefore, be expected to feature as excited states. Referring to the ^{17}O energy level diagram in Fig. 2.4, the first excited state is indeed $\frac{1}{2}^+$ and could therefore be interpreted as the $2s_{1/2}$ shell model state and there is also a $\frac{3}{2}^+$ state at 5.22 MeV which could be the $1d_{3/2}$ state. Detailed nuclear reaction studies support this interpretation which also indicates a $1d_{5/2}-1d_{3/2}$ spin–orbit splitting of $\simeq 5$ MeV in agreement with eq. (4.6). However, the plethora of other states indicates that the ^{16}O core must break up to form them. Of course, for even–even nuclei (and the ^{16}O core is one such) in which all nucleons are paired off in the ground state, excited states inevitably involve the breaking up of at least one pair and the situation becomes very complicated.

4.2.5 Comment

The foregoing description of the nucleus is successful in many ways. It is able to account for the magic numbers, the spins and parities of virtually all nuclear ground states, the general trend of nuclear magnetic moments and some excited states of simple nuclei having the structure 'closed shells $\pm$ 1 nucleon'. Further, it has been found to be an excellent starting point for more detailed calculations of the properties of nuclei reasonably close to a closed shell structure (section 4.4).

An implicit assumption made in using this model is that the nucleons move around fairly independently of each other. This may seem unrealistic given the very strong force between them which, on close approach, would be expected to scatter them continually from one shell model level to another. However, for such scattering to take place, there must be a vacancy (Pauli exclusion principle) in a level to which a nucleon is to be scattered. But for nucleons below the last filled level this is not the case since all lower shell model levels are filled. Scattering, therefore, cannot easily take place except in the region of unfilled shells and it is precisely this scattering which is taken into account in the more detailed calculations just referred to. Thus **independent motion** is a reasonable description of the behaviour of most nucleons in a nucleus.

4.3 THE COLLECTIVE MODEL

In the single-particle shell model just discussed the odd nucleon is treated as moving within a spherically symmetrical potential well due to the other nucleons. Of course the orbit of the odd nucleon itself can be highly asymmetrical depending on the value of l and inevitably, because of its interaction with the other nucleons, there will be some distortion of the core from spherical symmetry. The further the nucleus is from the very stable closed shell configurations the larger this distortion effect is likely to be and it is therefore perhaps not surprising that large quadrupole moments, indicating significant distortion of the whole nucleus, are observed in just these regions (section 4.2.3). This collective (involving many nucleons) effect is also what might be expected given that it has already been shown in section 4.1 that the nucleus does behave in some respects like a liquid drop.

The possibility of this sort of behaviour led Rainwater, Aage Bohr (son of Niels Bohr) and Mottelson in the early 1950s to introduce the collective model of the nucleus. It is best illustrated by considering the case of even–even nuclei.

Fairly near to a closed shell such a nucleus is expected to have spherical symmetry in its ground state. However, this spherical nucleus will be deformable and excited states are to be expected in which the nucleus oscillates about its spherical shape. For a fixed centre of mass the simplest possibility involving aspherical distortion is for the nucleus to undergo 'quadrupole' oscillations in which its shape varies between spherical and ellipsoidal. Since we are dealing with a quantum system these oscillations will be quantized with energy $\hbar\omega$ (say) where ω is the angular frequency of oscillation. These quantized oscillations are generally referred to as 'phonons' and, since the radius of an ellipsoidal shape as a function of θ is given by an expression of the form $r = A + BY_{20}(\theta)$, each phonon has an angular momentum quantum number $l = 2$ and even parity. Nuclei accurately described by this model should therefore have equally spaced (as with all oscillator quantum excitations) low lying excited states corresponding to the presence of $1, 2, \ldots$ phonons as illustrated in Fig. 4.5a. Note that with two phonons, each with $l = 2$, symmetry restricts the total angular momentum quantum number to 0, 2 or 4. In Fig. 4.5b the level diagram for a typical 'vibrational' nucleus (^{114}Cd) is given which has the expected features although the two-phonon states are not quite degenerate as predicted.

Another form of collective oscillation that can take place is one in which the protons and neutrons is a nucleus vibrate in antiphase with one another so as to form an oscillating electric dipole. The resultant state, known as a **giant dipole resonance**, has been identified in many nuclei through nuclear reaction studies. It occurs at very high

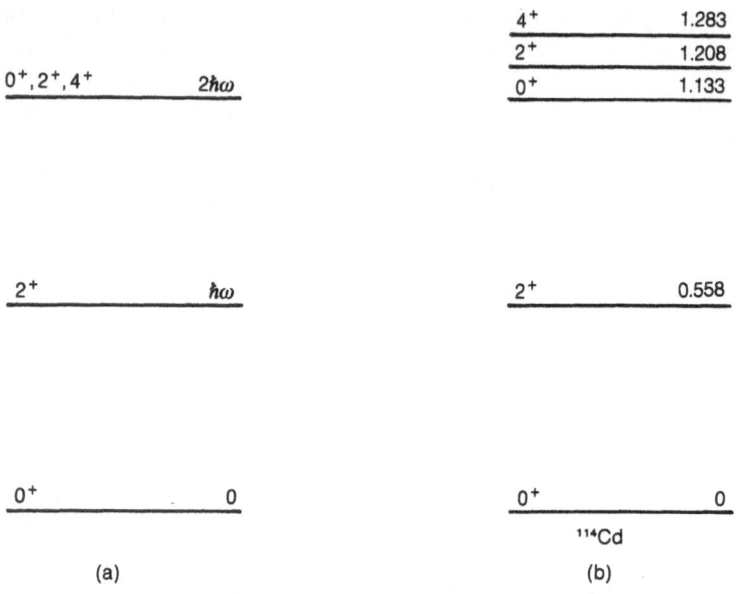

Fig. 4.5 (a) Theoretical level spacing of quadrupole vibrational levels; (b) low lying energy levels of ^{114}Cd.

energies, of the order 10–25 MeV depending on the nucleus, and has spin (parity) 1^-. In shell model terms it corresponds to the excitation of nucleons between different major shells – hence the high energy and odd parity.

Further away from closed shells the distorting effect of 'loose' or

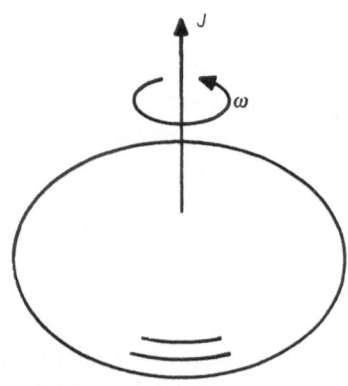

Fig. 4.6 Nuclear rotational motion with angular velocity ω.

'valence' nucleons can be such that the equilibrium shape of a nucleus is no longer spherical and the nucleus is permanently ellipsoidal – either prolate or oblate (Fig. 2.6). For such a distorted nucleus rotational motion is then to be expected about the minor axis of the ellipsoid (Fig. 4.6). The motion will again be quantized and the usual expression for the energy of rotational motion applies:

$$E_J = \frac{(\text{angular momentum})^2}{2I} = \frac{J(J+1)\hbar^2}{2I} \qquad (4.13)$$

where I is the moment of inertia associated with the motion. Because of the ellipsoidal symmetry J is restricted to the values $0, 2, 4, 6, \ldots$ for even–even nuclei and so for 'rotational' nuclei a simple sequence of levels ($J^P = 0^+, 2^+, 4^+, \ldots$) is predicted with energies given by eq. (4.3). This is frequently observed and Fig. 4.7 shows the low lying rotational levels for ^{238}U.

From the spacing of the rotational levels it is clearly possible to determine the values of the moments of inertia, I, for different nuclei. It is found that, in general, they are significantly smaller (by factors in the range 2–3) than the values for rigid rotation, implying that the rotational motion is somewhat fluid-like. This can be shown to result from the pairing force (section 4.2) which leads to nucleons pairing off in orbits with opposite angular momentum (j, m and $j, -m$). Such motion is completely different from rigid rotation in which all nucleons move together.

From the values of quadrupole moments and detailed studies of the rotational states and electromagnetic transitions between them (section 6.4) it emerges that, whereas oblate-shaped nuclei have relatively small asymmetries, prolate-shaped nuclei have major-to-minor axis ratios of up to 1.5:1 and in a one case (^{152}Dy – a **superdeformed** nucleus) of 2:1! For this same nucleus rotational states have been identified with J up to 60 (section 5.9). This is near to the limit (estimated as $J \simeq 80$) beyond which a nucleus would be expected to disintegrate because of the huge centrifugal force generated by the high angular momentum.

For such states it is also found that the moment of inertia approaches that for rigid rotation. Here it is interesting to plot the moment of inertia, I, as a function of the angular velocity of rotation, ω, for the state of lowest energy for a given angular momentum J. Such states are known as **yrast** states and a plot is shown in Fig. 4.8 for ^{156}Dy. It can be seen that, beyond a certain value of ω, I increases rapidly. This phenomenon, referred to as **back bending**, can be

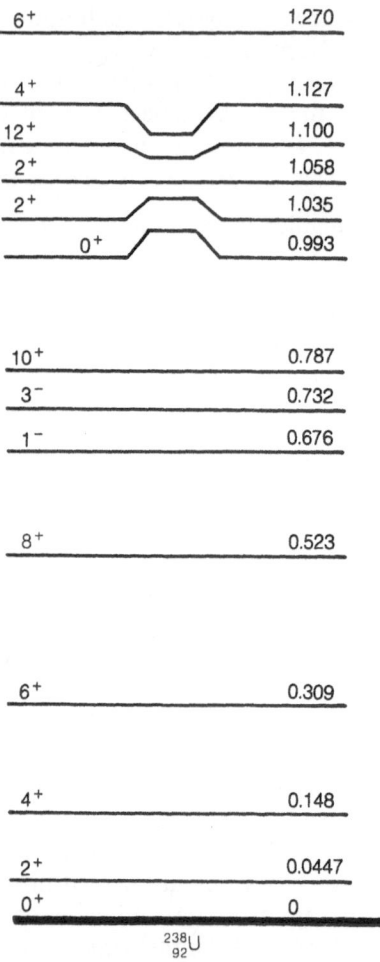

Fig. 4.7 Rotational energy levels of ^{238}U.

attributed to the Coriolis force ($\propto \omega$) which, at high angular velocities, overwhelms the pairing force which, as remarked earlier, is responsible for the lower values of I.

The foregoing discussion illustrates the nature of collective motion for even–even nuclei. For odd-A nuclei the situation is much more complicated and involves coupling the motion of the odd nucleon(s) to that of the vibrating or rotating core. Further, in the case of permanently distorted cores the motion of the nucleons has to be

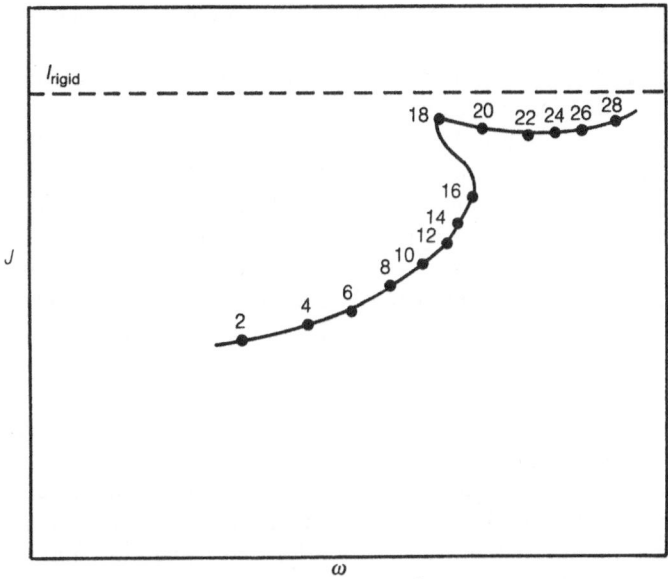

Fig. 4.8 Plot of I against ω for yrast states in ^{156}Dy.

evaluated in an ellipsoidal potential well which introduces further complications. Taking all these ramifications into account, however, has given a good understanding of the structure of many nuclei throughout the periodic table.

4.4 DEVELOPMENT OF NUCLEAR MODELS

The single-particle shell model and the collective model of even–even nuclei are two extreme representations of nuclear structure, neither of which gives a complete or accurate description. Ways in which they can be improved have already been briefly indicated.

In the case of the shell model many calculations involving all valence nucleons outside closed shells and taking account of the residual interaction between them have been carried out. They have been particularly successful in accounting for the properties of nuclei in the regions $4 < A < 16$ (filling of the 1p shell) and $16 < A < 40$ (filling of the 1d, 2s shells). This is difficult work involving many parameters to describe the residual interaction and requiring the diagonalization of matrices having thousands of terms. Interestingly, these calculations can reproduce collective phenomena but starting, of course, from the particle point of view, so the two models meet in this framework.

As has already been mentioned, calculations starting from the collective model but taking into account the motion of 'valence' nucleons in an ellipsoidal potential will enable some understanding of nuclear properties in regions of large distortion to be achieved.

More recently, because of the complexity of dealing with many individual nucleons, an approach has been developed (Arima and Iachello, 1975) in which nucleon pairs are treated as bosons (integer spin particles). This is known as the interacting boson model (IBM) and the bosons, which can represent excitations of closed shells, are taken to have spin–parity $J^P = 0^+$ or 2^+, the two lowest possible states for a pair of nucleons. One- and two-boson interactions are introduced and calculations can then proceed more simply and with fewer parameters than with the conventional shell model. The IBM cannot match the accuracy of shell model calculations (where they can be carried out), but the important thing is that it can be used in regions where shell model calculations are just too difficult, in particular the regions between spherical and strongly deformed nuclei. The model is now one of the main vehicles for understanding the structure of many nuclei.

5

Nuclear reactions

Much reference has been made in the last chapter to nuclear energy levels and their various properties (e.g. spins and parities) and we now move on to consider how information about these levels is obtained experimentally. There are two main ways of doing this – the use of nuclear reactions, and studies of how excited nuclei decay through the emission of β- and γ-radiation. In this chapter we deal with reaction processes.

A nuclear reaction is initiated by bombarding target nuclei with a beam of nucleons or nuclei. Although in the early days of nuclear physics it was common to use a beam of α-particles from radioactive decay (see Chadwick's neutron production experiment; section 1.4), nowadays the beam of particles comes from some sort of particle accelerator (section 5.4). As time has gone on it has become technically feasible to accelerate heavier and heavier nuclei, in the form of charged ions, and in the last decade or so many nuclear reaction studies have involved beams of heavy ions such as ^{16}O or ^{48}Ca.

In a nuclear reaction the collision process between target nuclei and beam particles leads to a rearrangement of the component nucleons, generally producing different product nuclei and particles. There are several types of reaction processes that can take place and these will be discussed in section 5.4 *et seq.* But consideration has first to be given to the following general aspects of all reaction processes:

1. energetics;
2. reaction probabilities as a function of energy, measured by the reaction cross-section;
3. angular distribution of the reaction products, measured by the differential cross-section.

These different aspects of nuclear reaction processes are discussed in the following sections.

5.1 ENERGETICS

Reference has already been made to the nuclear reaction studied by Chadwick, namely

$$^{9}_{4}\text{Be} + {}^{4}_{2}\text{He} \rightarrow {}^{12}_{6}\text{C} + \text{n} \tag{5.1}$$

which is conventionally written in the form $^{9}\text{Be}(\alpha, \text{n})^{12}\text{C}$. Consider now a general nuclear reaction denoted symbolically by

$$\text{A} + \text{a} \rightarrow \text{B} + \text{b}$$

or, simply, A(a, b)B. Here a is taken to be the bombarding particle (which might be a heavy ion) and A the target nucleus, B the residual nucleus and b an outgoing particle or nucleus which is detected. Of course the situation can be more complicated and there can be more than two reaction products but consideration here will be restricted to the simplest situation. The process will be governed by the usual conservation laws of energy, momentum and angular momentum. In addition, charge and the total number of nucleons will also be conserved.

From the energy point of view a critical quantity in considering the reaction is its Q value (nothing to do with the electric quadrupole moment for which, conventionally, the same notation is used). This is the difference in mass energy (mc^2) between the initial and final states, i.e.

$$Q = [(M_A + M_a) - (M_B + M_b)]c^2 \tag{5.2}$$

An alternative definition of Q can be obtained by using the conservation of energy during the reaction process. Thus, if the kinetic energies in the laboratory system of the different particles and nuclei taking part in the reaction are denoted by T_i (for the target nucleus A, being at rest, $T_A = 0$), conservation of energy requires

$$M_A c^2 + M_a c^2 + T_a = M_B c^2 + T_B + M_b c^2 + T_b \tag{5.3}$$

Combining eqs (5.2) and (5.3) then gives

$$Q = T_B + T_b - T_a \tag{5.4}$$

Q values can be positive (**exothermic** reaction) or negative (**endothermic** reaction) and will be zero in the case of an elastic scattering process of the type

$$\text{A} + \text{a} \rightarrow \text{A} + \text{a}$$

They are generally quoted in MeV and can be determined from eq. (5.2) using either nuclear masses or atomic masses since the number of electrons included in each pair of parentheses is the same. For the reaction given in eq. (5.1)

$$M(^4\text{He}) = 4.002\,60\,\text{u} \qquad M(^9\text{Be}) = 9.012\,19\,\text{u}$$
$$M(^{12}\text{C}) = 12.000\,00\,\text{u} \qquad M(\text{n}) = 1.008\,67\,\text{u}$$

so that

$$Q = 6.12 \times 10^{-3}\,\text{u} \times c^2$$

and using $1\,\text{u} \times c^2 = 931.19\,\text{MeV}$ (eq. (1.1)) this gives

$$Q = +5.7\,\text{MeV}$$

Values of Q for a few other typical reactions are as follows:

$$^2\text{D}(^2\text{D}, \text{n})^3\text{He} \qquad\qquad Q = 3.3\,\text{MeV}$$
$$^{12}\text{B}(\text{p}, \text{n})^{12}\text{C} \qquad\qquad Q = 12.6\,\text{MeV}$$
$$^{12}\text{B}(\text{p}, {}^3\text{H})^{10}\text{B} \qquad\qquad Q = -6.3\,\text{MeV}$$
$$^{27}\text{Al}(\gamma, \text{p})^{26}\text{Mg} \qquad\qquad Q = -8.3\,\text{MeV}$$

5.1.1 Centre of Mass System

Clearly, if Q is negative, energy conservation requires that kinetic energy must be brought in by the bombarding particle if the reaction is to take place. In the laboratory system this energy (**threshold** kinetic energy) must be larger than $|Q|$ because conservation of momentum does not allow the residual nucleus and the outgoing particle to be at rest in the laboratory. The situation becomes clear in the centre-of-mass system.

Denoting the speeds of a and A in this system by v_a and v_A respectively, conservation of momentum requires

$$m_a v_a + m_A v_A = 0$$

giving

$$v_A = -\frac{m_a}{m_A} v_a \qquad\qquad (5.5)$$

In the centre-of-mass system the total kinetic energy is (using eq. (5.5))

$$T = \tfrac{1}{2} m_a v_a^2 + \tfrac{1}{2} m_A v_A^2$$

$$= \tfrac{1}{2} m_a v_a^2 \left(1 + \frac{m_a}{m_A}\right) \qquad\qquad (5.6)$$

In the laboratory system, remembering that the target nucleus is at rest, the total kinetic energy is

$$T' = \tfrac{1}{2}m_a v_a'^2 \tag{5.7}$$

where v_a' is now the speed of a in this system. But v_a' is simply the relative velocity of a and A, i.e.

$$v_a' = v_a - v_A$$

Substituting this expression for v_a' in eq. (5.7) and using eq. (5.5) then gives

$$T' = \tfrac{1}{2}m_a v_a^2 \left(1 + \frac{m_a}{m_A}\right)^2 \tag{5.8}$$

Comparing eqs (5.6) and (5.8) we have finally

$$T' = T\left(1 + \frac{m_a}{m_A}\right) \tag{5.9}$$

The condition for an endothermic reaction to proceed requires that $T \geqslant |Q|$ or, using eq. (5.9),

$$T' \geqslant |Q|\left(1 + \frac{m_a}{m_A}\right) \tag{5.10}$$

Clearly, from the point of view of accelerator energy needed, it would be advantageous if the laboratory and centre-of-mass systems coincided. This would require bombarding particle and target to approach each other with the same momentum and we shall see (section 5.4 and Chapter 9) that this approach ('colliding beam' experiment) is sometimes used in particle physics. Of course, in particle physics accelerator energies are very high and relativistic mechanics must be used in order to obtain relationships between laboratory and centre-of-mass energies.

5.2 NUCLEAR REACTION CROSS-SECTION

In discussing the probability of a given nuclear reaction's taking place it is usual to introduce the concept of a **cross-section**. This is an easily understood classical concept. Consider a beam of point particles impinging on a target consisting of a collection of spherical objects of radius R (Fig. 5.1a). Each target sphere presents a cross-sectional area $\sigma = \pi R^2$ to the beam and, considering a single sphere (Fig. 5.1b), all particles arriving within that area will strike it. If the incident particles themselves are spheres of radius r (say) then the area of interaction

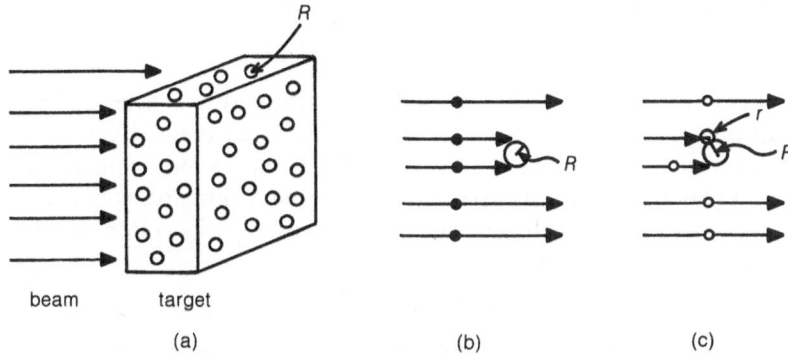

Fig. 5.1 (a) Interaction of a beam of particles with a target consisting of spheres; (b) point particles and (c) particles of radius r interacting with a sphere of radius R.

would of course be increased to $\sigma = \pi(r + R)^2$ (Fig. 5.1c); this is the same mathematically as if point particles were striking spheres of radius $(r + R)$. Clearly the probability of a collision increases proportionally to σ.

Turning now to a nuclear reaction, we no longer have well-defined spheres for target nucleus or incident particle since their densities vary with radius (Fig. 2.1) and they have essentially 'fuzzy' edges. In addition, because of the finite range of nuclear forces, there can be an interaction when the matter distributions of the colliding bodies are not in direct contact. Nevertheless, the probability of a reaction's taking place can still be measured in terms of a notional cross-sectional area which we shall continue to denote by σ. σ can refer to the **total** cross-section (denoted by σ_T) which relates to the probability of anything happening when particle and target interact or it can be a **partial** cross-section σ_i which relates to the probability that a specific reaction (denoted by i) will take place. Clearly

$$\sigma_T = \sum \sigma_i \qquad (5.11)$$

Although σ is not the geometrical cross-section of a nucleus, even if that could be precisely defined, and can be larger or smaller than this, it is nevertheless to be expected that it will be of that order of magnitude. Since nuclei have radii R in the region from 2 fm to 7 fm (section 2.1), so that πR^2 is in the region from 5×10^{-29} m^2 to 1.5×10^{-27} m^2, it is usual to express nuclear reaction cross-sections in units of the **barn**:

$$1 \text{ barn} = 1 \text{ b} = 10^{-28} \text{ m}^2 \tag{5.12}$$

Consider now how the attenuation of a beam of particles passing through a target is related to σ. This process is represented in Fig. 5.2 in which a target of area A, thickness dz and containing n nuclei per unit volume is struck by a beam consisting of N bombarding particles per second striking the area A. If the reaction cross-section of each nucleus is σ, we have

$$\text{total number of nuclei in target} = nA\,dz$$

aggregate cross-section presented by nuclei in target $= n\sigma A\,dz$
total number of bombarding particles interacting with a nucleus
= (number striking unit area) × (aggregate cross-section)

$$= \frac{N}{A} n\sigma A\,dz$$

Denoting the magnitude of the last quantity by dN then gives the following expression for the number of particles removed (by

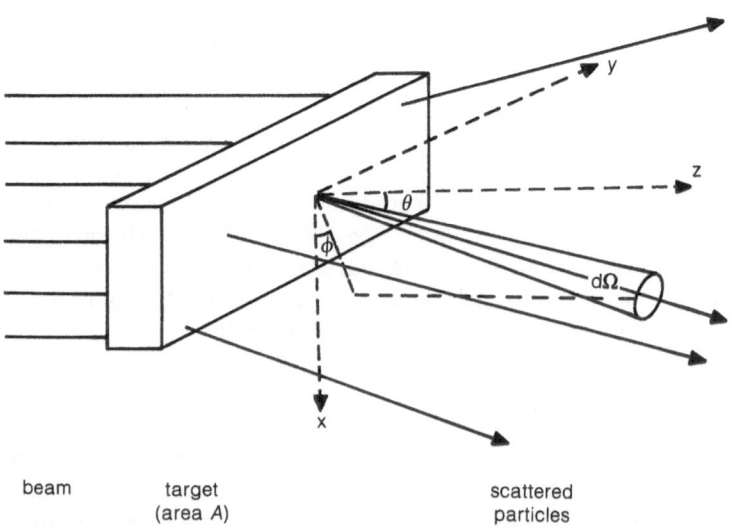

Fig. 5.2 Particle beam striking a target of area A resulting in scattered particles.

interaction) from the beam

$$\frac{dN}{N} = -n\sigma \, dz \tag{5.13}$$

the minus sign indicating removal. For a finite target of thickness z, the attenuation of the beam is simply obtained by integrating eq. (5.13) over the distance z, i.e.

$$\int_{N_0}^{N} \frac{dN}{N} = \int_{0}^{z} -n\sigma \, dz \tag{5.14}$$

where N_0 refers to the intensity of the incident beam and N that of the transmitted beam. Integrating eq. (5.14) gives

$$N = N_0 e^{-n\sigma z} = N_0 e^{-\mu z} \tag{5.15}$$

where $\mu = n\sigma$ is known as the **linear attenuation coefficient**. Its inverse, $1/n\sigma$, which has the dimensions of a length, is the distance over which the beam intensity is reduced by a factor $1/e$; it is also the mean free path between collisions.

Finally we introduce the concept of **differential cross-section** which is denoted by $d\sigma/d\Omega$. This is used when the angular distribution of the products of a nuclear interaction is under consideration. The situation is illustrated in Fig. 5.2 where reaction products are shown moving in a direction specified by the usual spherical polar angles θ and ϕ into the solid angle $d\Omega = \sin\theta \, d\theta \, d\phi$. For a given nuclear reaction, whilst σ relates to the probability of the reaction taking place and the reaction products moving in all possible directions, $d\sigma/d\Omega$ relates to the probability of the reaction products' being found moving in the direction (θ, ϕ). $d\sigma/d\Omega$ will in general be a function of θ and ϕ although, if the bombarding particles and target nuclei have their spins randomly oriented, the direction of incidence will be an axis of symmetry and $d\sigma/d\Omega$ will be independent of ϕ. If the differential cross-section is integrated over the full solid angle 4π the cross-section σ is obtained: thus

$$\sigma = \int^{4\pi} \frac{d\sigma}{d\Omega} d\Omega \tag{5.16}$$

or, if $d\sigma/d\Omega$ is independent of ϕ,

$$\sigma = 2\pi \int_{0}^{\pi} \frac{d\sigma}{d\Omega} \sin\theta \, d\theta \tag{5.17}$$

Differential cross-sections are inevitably much smaller than total cross-sections and are generally measured in millibarns per steradian (mb sr^{-1}).

5.3 THE EXPERIMENTAL APPROACH TO NUCLEAR REACTIONS

Having defined the quantities that are normally measured in a nuclear reaction we here outline the typical experimental procedures which are followed for studying the symbolic reaction $A(a, b)B$. No details are given of the apparatus other than to mention very briefly the underlying physical principles. Details of low energy nuclear physics apparatus are given, for example, in Burcham (1988) and of high energy elementary particle apparatus in a book in this series by Kenyon (1988).

Referring to Fig. 5.3, charged ions of the particle a are produced in some form of accelerator (described later in this section) and, by use of bending magnets for example, will emerge with a particular energy. These ions then pass through a collimator in order to define their direction with some precision and strike a target containing the nuclei A. As the beam particles move through the target they will mainly lose energy by ionizing target atoms and so, if precise energy measurements are to be made, a thin target must be used. This, however, increases the difficulty of the experiment since few interactions will take place. Choice of target thickness is clearly a crucial decision in planning an experiment.

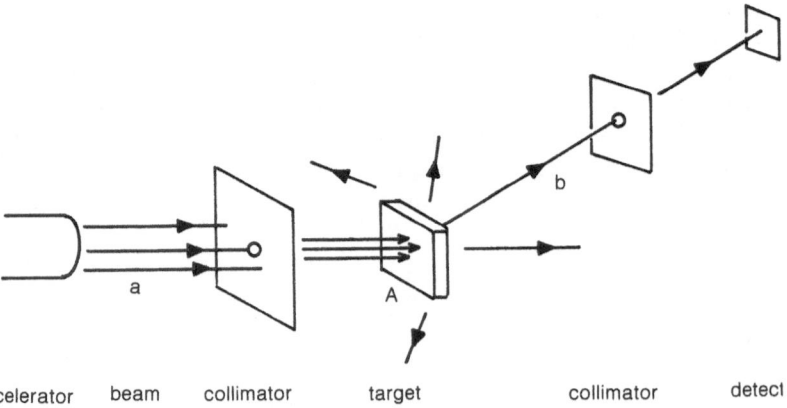

accelerator   beam   collimator   target   collimator   detect

Fig. 5.3 Layout of apparatus for studying the reaction $A(a, b)B$.

The reaction product particles b move off in all directions and their angular distribution can be studied by detecting them after passage through another collimator set at a particular angle θ. Various types of detector are used (discussed later) – sometimes in combination – and these can determine the type of particle as well as its energy. But experimenters have to contend with many complications of interpretation, impurities in targets and, not least, the stability of their apparatus. In the end, detailed information becomes available about σ, $d\sigma/d\Omega$ and their energy dependence for the reaction under study.

5.3.1 Accelerators

Most important for nuclear reaction studies are Van de Graaff accelerators in which ions are accelerated in an evacuated tube by an electrostatic field maintained between a high voltage terminal and an earth terminal, charge being conveyed to the high voltage terminal by a rotating belt or chain. In early forms of this accelerator, positive ions from a gaseous discharge tube were accelerated from the high voltage terminal to earth. But, in modern 'tandem' accelerators, negative ions are accelerated from earth to the high voltage terminal where they are then stripped of some electrons and the resultant positive ions are further accelerated down to earth potential. The effective accelerating potential is thus twice the potential difference in the machine. High flux proton beams with energies up to around 30 MeV can be produced in this way. The machines can also be used to accelerate heavy ions such as ^{16}O.

At higher energies use is generally made of orbital accelerators in which charged particles are confined to move in circular orbits by a magnetic field. At non-relativistic energies the angular frequency of rotation ω, known as the **cyclotron frequency**, is constant depending only on the strength of the field. In a **cyclotron**, the particles rotate in a circular metallic box split into two halves, known as Ds, between which an oscillating electric field is maintained. Its frequency matches ω and so the particle is continually accelerated. In a fixed magnetic field the orbital radius increases as the energy increases and, at some maximum radius, the particles are extracted using an electrostatic deflecting field. However, as the energy becomes relativistic (remember $m_p \simeq 1000 \, \text{MeV}/c^2$), ω decreases with energy and it becomes necessary to decrease steadily the frequency of the oscillating electric field with energy to preserve synchronization.

Such a machine is known as a **synchrocyclotron** and protons with energies in the region of 100 MeV have been produced in this way.

For energies higher than this gigantic magnets would be needed and so the approach is to accelerate bunches of particles in orbits of essentially constant radius using annular magnets producing magnetic fields which increase as the particle energy increases. This energy increase is provided by passing the particles through radio-frequency cavities whose frequency also changes slightly as the particles are accelerated to ensure synchronization. Such devices are called **synchrotrons** and can be physically very large. For example, the so-called Super Proton Synchrotron (SPS) at CERN (Geneva) has a circumference around 6 km and can produce protons with energies up to around 450 GeV. LEP (the Large Electron–Positron Collider) has a circumference of 27 km and accelerates electrons (and positrons in the opposite direction) to energies of $\simeq 60$ GeV or more. Finally, the Superconducting Super Collider (SCC), which uses superconducting magnets, and which is being built in the USA, has a circumference of 87 km and will produce proton and antiproton beams with energies $\simeq 20\,000$ GeV!

Electrons can also be accelerated in synchrotrons but, because of their small mass, large amounts of energy are radiated (synchrotron radiation) owing to the circular acceleration. At energies beyond a few GeV this loss becomes prohibitive and use has to be made of linear accelerators in which electrons are accelerated down a long evacuated tube by a travelling electromagnetic wave. The Stanford Linear Accelerator (SLAC) in the USA, for example, is around 3 km long and can produce pulses of electrons with energies up to 50 GeV.

5.3.2 Detectors

Although in the early days much use was made of ionization chambers, for example the Geiger counter (section 1.4), the detectors currently in use for nuclear physics experiments are usually either **scintillation counters** or **semiconductor detectors** or some combination. The former are developments of the approach of Rutherford, Geiger and Marsden (section 1.3) using the scintillations produced in a ZnS screen to detect α-particles. Various scintillators are in current use such as NaI activated by an impurity (usually thallium for detection of γ-particles), or some organic material dissolved in a transparent plastic or liquid. The scintillations are detected by a photomultiplier tube producing a pulse of photoelectrons. The size of the pulse – the pulse height – gives a measure of the energy of the incident particle.

Semiconductor detectors depend on an incident particle or photon exciting an electron from the valence band to the conduction band. The resultant increase in conductivity – a conduction pulse – then produces a signal which is processed electronically and which enables the energy of the incident radiation to be measured.

In the field of very high energy physics, considerable use is made of **bubble chambers** and **wire chambers**. The former follows on from the Wilson cloud chamber and consists essentially of a large chamber, possibly several metres in diameter, containing liquid (e.g. hydrogen, helium, propane, ...) near its boiling point. The chamber is expanded as charged particles pass through it, leading to the formation of bubbles, as a result of boiling, along the particle tracks which can be stereo flash photographed. The lengths of the tracks and their curvature in a magnetic field enable particle lifetimes, masses and energies to be deduced.

Wire chambers consist of stacks of positively and negatively charged wire grids in a low pressure gas. An incident charged particle ionizes the gas and acceleration of the resultant electrons near the anode wires leads to further ionization and an electrical pulse. The physical location of the pulse can be determined electronically so that track measurements can be made. Using an applied magnetic field to bend the tracks again enables information to be obtained about the properties of the detected particle.

5.4 NUCLEAR REACTION PROCESSES

In the previous chapter some understanding of nuclear structure has been achieved in terms of a nuclear model in which nucleons move around fairly independently in a potential well. To give some intuitive understanding of nuclear reaction processes we stay with this description of the nucleus and follow a very illuminating discussion given by Weisskopf (1957).

An incident particle, a, approaching a nucleus, A, will, if it is charged, first experience the long-range Coulomb potential and if its energy is low will be elastically scattered by this before coming within the range of the nuclear force. In this case it undergoes **Rutherford** or **Coulomb scattering** as described in section 1.3. For higher energies, or if the particle is uncharged, it will come within the range of the nuclear force due to the nucleons. This can be represented by a potential energy curve of the form already discussed and shown in Fig. 4.2 for an incident neutron or proton. As a result of this interaction the incident particle may again be elastically scattered without colliding directly with a nucleon in the nucleus. The form of this scattering will obviously depend on the shape and size of the nucleus and its associated potential well, and is referred to as **shape elastic scattering**. All of these processes are symbolized by

$$A + a \rightarrow A + a$$

If a direct collision with a nucleon takes place then there are various possibilities. The nucleon may be excited to a higher (unoccupied) state and the incident particle leaves the nucleus with reduced energy. This is an **inelastic scattering** process and the nucleus is left in an excited state. Another variant of this is that, instead of exciting a nucleon, the incident particle excites a collective mode – a vibrational or a rotational state (section 4.3). Such processes are symbolized by

$$A + a \rightarrow A^* + a$$

where A^* signifies an excited state of the nucleus A.

Alternatively the incident particle may give enough energy to the nucleon with which it collides so that this nucleon, b, is knocked out from the nucleus (Fig. 5.4a). There are two possibilities here, depending on how much energy is lost by the incident particle. If the incident particle, a, retains enough energy to escape from the nucleus after the collision we have the process

$$A + a \rightarrow B + a + b$$

where B is the residual nucleus remaining after the nucleon b has been knocked out of A. However, the incident particle may lose so much energy that it is captured, resulting in the formation of a nucleus B', i.e.

$$A + a \rightarrow B' + b$$

Reactions of the various kinds just discussed are referred to as **direct reactions** since there is direct interaction with a single nucleon rather than with the nucleus as a whole. Other variants illustrated in Figs 5.4b and 5.4c are **stripping** and **pick-up reactions**. In the former a composite incident particle, usually a deuteron (^{2}H), is stripped of one of its component nucleons which remains in the target nucleus and the remaining nucleon(s) escape. Conversely, in the latter, the incident

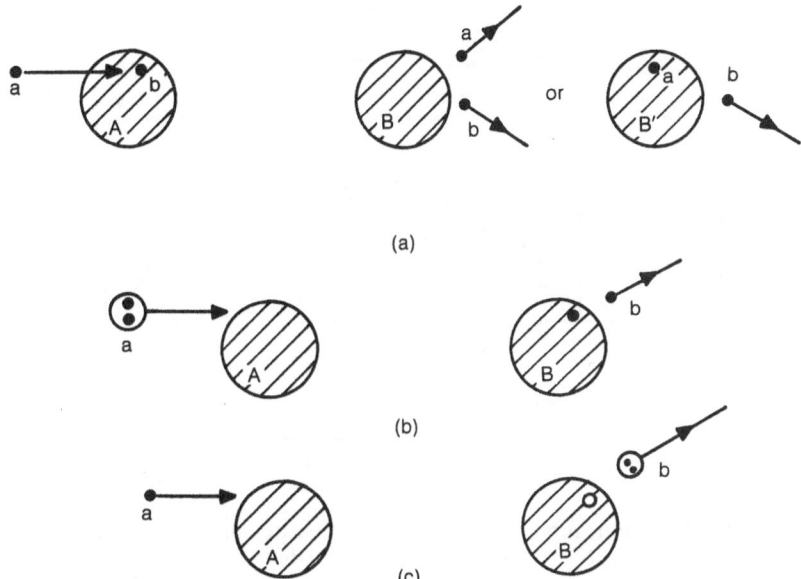

Fig. 5.4 Examples of direct nuclear reaction processes: (a) 'knock-out', (b) 'stripping' and (c) 'pick-up'.

particle, usually a nucleon, picks up another nucleon from the target nucleus and carries it away, emerging as a deuteron.

The next possibility is that the incident particle collides with a nucleon in the target nucleus, perhaps one lying in a very low shell model level, and neither has sufficient energy to escape. There will then be a series of further random collisions in the nucleus (Fig. 5.5) until eventually enough energy is concentrated by chance on one particle to enable it to escape; or the nucleus may lose its energy by emitting electromagnetic radiation. This state of the nucleus after it has captured the incident particle and in which many internal collision processes are occurring was first discussed by Niels Bohr in 1936 and is referred to as the **compound nucleus**. Whereas a direct reaction, for an incident particle of several MeV, takes place in a time of the order of that taken by a nucleon to cross a nucleus ($\simeq R/c$ $\simeq 10^{-14}\,\text{m}/10^8\,\text{m s}^{-1} = 10^{-22}\,\text{s}$) the compound nucleus exists for a much longer period and we shall see in section 5.8 that it can exist for times in the approximate region from $10^{-14}\,\text{s}$ to $10^{-20}\,\text{s}$. A compound nucleus process can thus be represented as taking place in two stages – formation of the compound nucleus and, after a

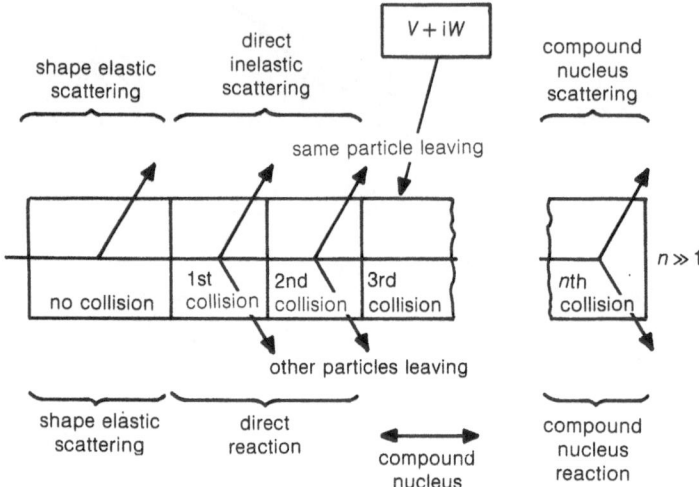

Fig. 5.5 Schematic representation of the various possible stages of a nuclear reaction in terms of the number of collisions which take place.

considerable time, its decay. Symbolically,

$$A + a \rightarrow C^* \rightarrow B + b$$

where C^* represents the excited compound nucleus. Because of the long life of the compound nucleus little information about its mode of formation is carried forward to influence the way in which it disintegrates.

Finally, one other approach to coping with the complexities of nuclear reaction processes should be mentioned, namely, the **optical model** which was developed by Feshbach, Porter and Weisskopf in 1954 and which is useful in giving a broad understanding of nuclear reactions. It is based on the simple idea that, considering only the behaviour of the incident particle, on average it experiences a potential well of the type used in shell model discussions but including an imaginary component and denoted by $V + iW$. The latter term is introduced to allow for the fact that the incident particle may be absorbed through some form of reaction. This is analogous to the imaginary term in the refractive index introduced in optics to take account of absorption. The nucleus is thus treated, and frequently referred to, as a **cloudy crystal ball**!

We now briefly elaborate the different aspects of nuclear reactions just outlined.

5.5 SCATTERING AND ABSORPTION

In this section we consider some general features of scattering resulting from the interaction of the incident particle (if it is charged) with the Coulomb field of the nucleus and with the average optical potential just mentioned. In the discussion the input ingredients are the electric charge of the nucleus, the average nuclear potential energy, $V(r)$, experienced by the incident particle when it gets close to the target nucleus, and the fact that it may be absorbed through the initiation of some form of nuclear reaction. The absorption is represented on average by the imaginary part, $iW(r)$, of the optical potential. No consideration is given of the details of possible reactions and attention is concentrated simply on the average behaviour of the incident particle.

5.5.1 Rutherford (Coulomb) Scattering

We consider first the scattering of a charged particle in the Coulomb field of a nucleus, referred to as **Rutherford** or **Coulomb scattering**. A particle of charge ze approaching a nucleus of charge Ze will, at large distances, experience only the Coulomb potential

$$V_C(r) = \frac{Zze^2}{4\pi\varepsilon_0 r} \tag{5.18}$$

at distance r from the centre of the nucleus. When it comes within range of the attractive nuclear force this potential is radically changed as shown symbolically in Fig. 5.6. Here the average effect of interactions through the nuclear force is represented by the optical model potential $V(r) + iW(r)$. There is clearly a Coulomb potential barrier V_B to be surmounted if the incident particle is to interact with this potential whose value is approximately

$$V_B \simeq \frac{zZe^2}{4\pi\varepsilon_0 R} \tag{5.19}$$

where R is the nuclear radius. Typical values for V_B in the case of an incident proton ($z = 1$) are

$$\text{carbon } (Z = 6): \quad V_B \simeq 3 \, \text{MeV}$$
$$\text{silver } (Z = 47): \quad V_B \simeq 12 \, \text{MeV}$$
$$\text{lead } (Z = 82): \quad V_B \simeq 17 \, \text{MeV}$$

Although, classically, an incident particle will only be able to

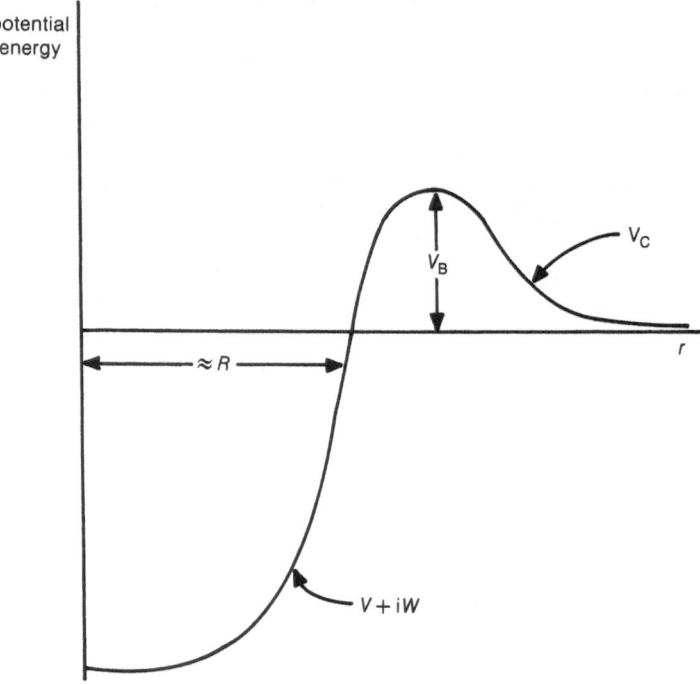

Fig. 5.6 Potential experienced by a charged particle in the vicinity of a nucleus.

surmount the barrier if its energy is greater than V_B this is not so quantum mechanically, and there is a finite chance of penetrating the barrier for energies less than V_B. However, if the energy is well below the barrier height then this effect is negligible and the incident particle will be simply elastically scattered by the Coulomb potential. For a 'head-on' collision under this circumstance, the classical distance of closest approach, d_0, is simply obtained by equating the kinetic energy, T, of the incident particle in the centre-of-mass system to the value of V_C for $r = d_0$. This gives

$$d_0 = \frac{zZe^2}{4\pi\varepsilon_0 T} \tag{5.20}$$

Coulomb scattering is represented diagramatically in Fig. 1.2 where θ is the angle of scattering. Detailed calculations show that the trajectory of the scattered particle is a hyperbola and that the probability of its being scattered through an angle θ expressed via the differential cross-section (section 5.2) is given by

$$\frac{d\sigma}{d\Omega} = \left(\frac{1}{4\pi\varepsilon_0}\right)^2 \left(\frac{zZe^2}{4T}\right)^2 \operatorname{cosec}^4\left(\frac{\theta}{2}\right)$$

$$= \frac{d_0^2}{16} \operatorname{cosec}^4\left(\frac{\theta}{2}\right) \tag{5.21}$$

The latter expression clearly shows that the differential cross-section has the dimensions of an area as required. A plot of $d\sigma/d\Omega$ against θ for the case of 7.68 MeV α-particles scattered by ^{197}Au, as in the original Rutherford, Geiger and Marsden experiment, is shown in Fig. 5.7. Here the finite probability of backward scattering ($\theta \simeq 180^0$),

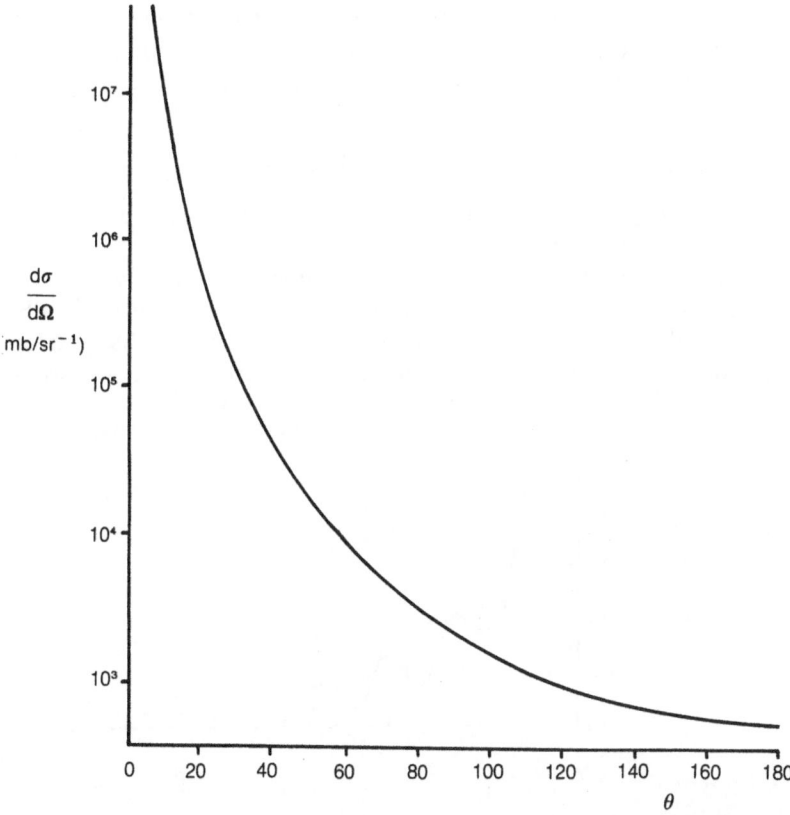

Fig. 5.7 Rutherford scattering of 7.68 MeV α-particles by ^{197}Au.

which indicated to Rutherford that the nuclear charge was concentrated into a small volume, can clearly be seen.

The foregoing discussion has been carried out assuming that the electrical charges behave as though they are concentrated at the centres of the incident particle and the scattering nucleus. This is an acceptable assumption provided that they do not approach each other too closely but at higher energies and when heavy ions (section 5.9) are being scattered modifications of the above expressions have to be incorporated.

5.5.2 Scattering and Absorption by the Nuclear Potential

Consider now the situation when the energy of the incident particle is sufficiently high that the distance of closest approach, d, brings it within range of the nuclear optical potential. d depends on the angle of scattering and is smallest, d_0, for a head-on collision for which the scattering angle is $\theta = 180^\circ$. In general, d is given by

$$d = \frac{d_0}{2}\left[1 + \operatorname{cosec}\left(\frac{\theta}{2}\right)\right] \tag{5.22}$$

which clearly decreases as the scattering angle increases. For higher energy charged particles $d\sigma/d\Omega$ is therefore expected to deviate from the expression given in eq. (5.21) for larger values of θ. This is illustrated in Fig. 5.8 for the scattering of 30 MeV protons by ^{208}Pb. The regular fluctuations for higher values of θ, which are observed for a wide range of nuclei and also for incident neutrons, are reminiscent

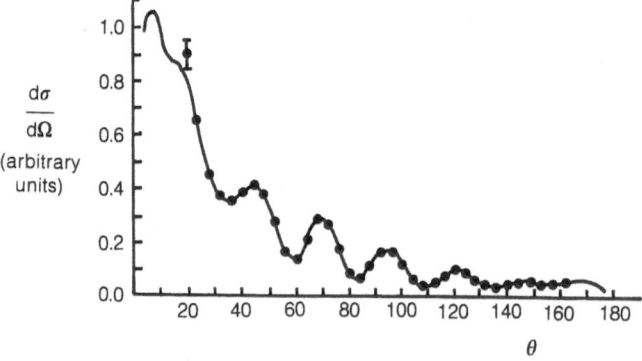

Fig. 5.8 The differential cross-section ($d\sigma/d\Omega$) for the scattering of 30 MeV protons by ^{208}Pb. Comparison between theory and experiment.

of a 'smeared-out' diffraction pattern and can indeed be interpreted as approximating to Fraunhofer diffraction due to an absorbent sphere. The absorption derives from the term iW in the optical potential and the separation of the peaks in the diffraction pattern is related to the radius of the sphere, giving values in the range $R = (1.4–1.5)A^{1/3}$ fm. As is to be expected this expression has the same dependence on A as the nuclear radius but is somewhat larger (eq. (2.2)) owing to the range of the nuclear force. In order to account for the scattering data for nucleons at a few tens of MeV, well depths for V and W of the order of 50 MeV and 5 MeV respectively are needed. The relatively small value of the latter reflects the inhibiting effect of the Pauli principle on collisions (discussed in section 4.2) which could lead to absorption of the incident particles. The nucleus is thus fairly 'transparent' to incident particles although there is enough absorption to lead to the observed diffraction scattering. The nucleus truly behaves as a 'cloudy crystal ball'.

Another feature accounted for by the optical model is the behaviour of the average total cross-section, $\bar{\sigma}_T$ (i.e. total cross-section for scattering and absorption averaged over any fluctuations due to small changes in energy) as a function of energy and as a function of nuclear size. It is found that $\bar{\sigma}_T$ is a smooth function of energy for each nucleus and that broad, shallow peaks occur. These can be understood as resonances (known as **size resonances**) occurring whenever, roughly speaking, a number of (de Broglie) wavelengths for the incident particle can be fitted into the potential well V. Since the scattering time is of the order of $\Delta t \simeq 10^{-22}$ s (section 5.4) there is an uncertainty in the energy of such a resonant state of the order of $\Delta E \simeq \hbar/\Delta t \simeq 10$ MeV. The state is said to have a **width** $\Gamma \simeq 10$ MeV; it is broad as just stated.

The optical model has been used successfully up to incident nucleon energies of several hundred MeV to account for the average behaviour of scattering, absorption and differential cross-sections. It is most successful at energies above $\simeq 10$ MeV and for medium weight and heavy nuclei.

5.6 COULOMB EXCITATION

Although a charged particle with energy well below the Coulomb barrier cannot interact directly with the target nucleus through the nuclear force, it can change the state of the nucleus through the electromagnetic force. This allows the possibility of what are known as **Coulomb excitation** processes. Such processes have been of importance, for example, in exciting nuclear rotational states (section 4.3) through interaction of the incident charged particle with the electric quadrupole moment of a nucleus. Heavy ions, because of their high charge, are particularly effective in this respect.

5.7 DIRECT REACTIONS

Direct reactions, as outlined in section 5.4, are characterized by a single collision of an incident particle with a nucleon in the nucleus. This collision normally takes place in the region of the nuclear surface since deeper penetration is likely to lead to multiple collisions and the formation of a compound nucleus. The nucleon may be excited (inelastic scattering), or picked up (e.g. in a (p, d) reaction), or a nucleon from an incident composite particle may be captured into a nuclear energy state (e.g. in a (d, n) reaction). In whatever form, the reaction generally leads to the population of a low excited state of the nucleus. Involving only one collision, such reactions are reasonably easy to understand and interpret and are particularly useful for testing shell model descriptions of nuclear states.

The main source of information is the form of the angular distribution, specified by $d\sigma/d\Omega$, of the outgoing particles. Consider Fig. 5.9 in which an incident particle of momentum p_i interacts directly with a nucleus resulting in the emission of an outgoing particle of momentum p_f moving at an angle θ to the direction of incidence. Momentum q is transferred to the nucleus at the point of

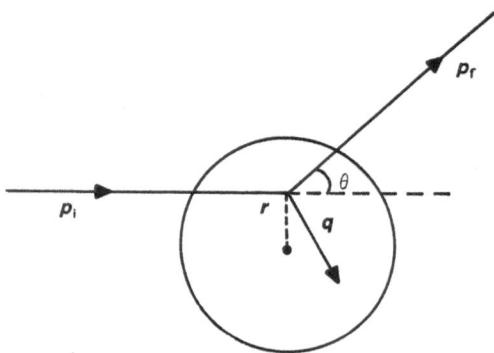

Fig. 5.9 A direct interaction process in which momentum $q(=p_i-p_f)$ is transferred to a nucleus.

collision r where $q = p_i - p_f$. The magnitude of q is dependent on θ and is given by the usual triangular relationship

$$q^2 = p_i^2 + p_f^2 - 2p_ip_f \cos \theta \qquad (5.23)$$

The angular momentum transferred to the nucleus is

$$L = r \times q \qquad (5.24)$$

and this will be quantized and have magnitude $[l(l+1)]^{1/2} \hbar$ where l is the usual orbital angular momentum quantum number. For excitation of a nucleus from its ground state to a particular excited state, angular momentum conservation severely restricts the value(s) of l that can contribute, and a further limitation is imposed by parity conservation which requires that the difference in parity between the

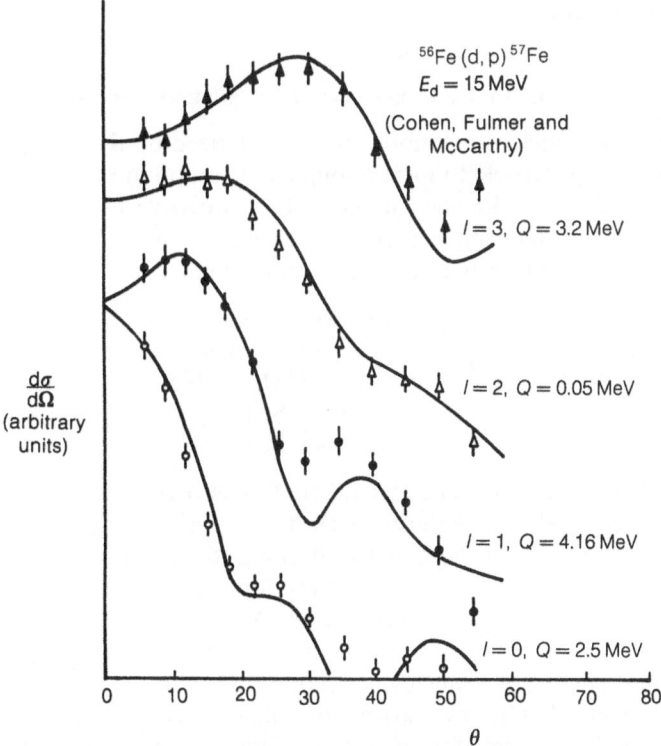

Fig. 5.10 Experimental plots of $d\sigma/d\Omega$ against scattering angle θ for various stripping processes in the reaction ^{56}Fe(d, p) ^{57}Fe.

two states must equal $(-1)^l$. Assuming that the collision takes place near the nuclear surface (i.e. $r \simeq R$), these limitations on l mean that the condition (5.24) is only satisfied at various points. Interference between outgoing particles coming from these points then leads to a diffraction-like pattern for the angular distribution. The larger is the angular momentum ($\simeq l\hbar$) transferred to the nucleus, the larger q must be and, therefore, the larger θ (eq. (5.23)). Thus the general expectation is that the first (main) peak of the diffraction pattern should occur at higher values of θ as the required value for l increases.

Typical experimental angular distributions for stripping reactions illustrating these features are shown in Fig. 5.10. Through detailed analysis of experimental results of this kind much information has been obtained about the spins and parities of the excited states produced by the direct interaction processes and the extent to which they can be adequately described by shell model or collective configurations.

5.8 COMPOUND NUCLEUS REACTIONS

Compound nucleus reactions are characterized by the capture of a bombarding particle to form a long-lived compound state in which many collisions take place followed by the decay of the compound states. A typical example is the capture of protons by ^{27}Al for which, *inter alia*, the following processes can take place

$$^{27}\text{Al} + \text{p} \rightarrow {}^{27}\text{Al} + \text{p}$$
$$\rightarrow {}^{27}\text{Al}^* + \text{p}$$
$$\rightarrow {}^{24}\text{Mg} + {}^{4}\text{He}$$
$$\rightarrow {}^{27}\text{Si} + \text{n}$$
$$\rightarrow {}^{28}\text{Si} + \gamma$$

The different decay modes are referred to as **channels** and, as can be seen, include elastic scattering, inelastic scattering, the production of different nuclei and the radiative decay of the compound state.

Measurements of the cross-section for reactions of this kind for incident nucleon energies up to a few MeV only, i.e. for excitation energies of the compound state in the region $\simeq 8$ MeV (average binding energy of a nucleon – section 2.2) plus a few MeV, show the presence of many narrow resonances. This is illustrated in Fig. 5.11 for the last of the above reactions. A resonance occurs when the energy of the incident particle is such that the resultant excitation energy of the compound nucleus corresponds to that of one of its

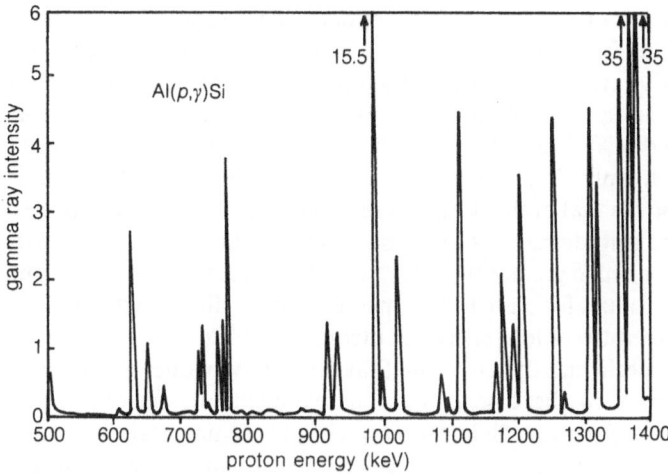

Fig. 5.11 Compound nucleus resonance in the reaction $^{27}\text{Al} + \text{p} \rightarrow {}^{28}\text{Si}^*$ $\rightarrow {}^{28}\text{Si} + \gamma$.

(quasi-stationary) excited states. Such states, having energies greater than 8 MeV or so, are extremely complicated and are relatively closely spaced compared with the separation of low-lying states.

Denoting the energy of an incident particle at resonance by E_r, it is found that, in the vicinity of the resonance, the cross-section (σ) for a particular nuclear reaction (i) depends on the energy (E) of the incident particle according to the following expression

$$\sigma(E) = \pi \lambdabar^2 g \frac{\Gamma_c \Gamma_i}{(E - E_r)^2 + \frac{1}{4}\Gamma^2} \qquad (5.25)$$

where $\lambdabar = \lambda/2\pi$, with λ the de Broglie wavelength of the incident particle, and g is a statistical factor depending on the spins of the particles involved. Γ is the width of a resonance peak at half-maximum. Γ_c is proportional to the probability for formation of the compound state via the initiating incident particle and target nucleus ($^{27}\text{Al} + \text{p}$ in the above example). Γ_i is proportional to the probability of the compound state's decaying via the channel i and the Γ_i are referred to as **partial widths**. Γ is the sum of all the Γ_i (including Γ_c, which is simply the width for the elastic channel), i.e.

$$\Gamma = \sum \Gamma_i \qquad (5.26)$$

The expression for the dependence of σ on energy is known as the Breit–Wigner formula and is exactly analogous to the expression used for a resonant electrical circuit when responding to an input of varying frequency. The form of its dependence on Γ_i implies that the relative cross-sections for different decay channels of the compound nucleus are independent of its mode of formation (**independence hypothesis**). The relative cross-sections are thus independent of Γ_c which only determines their absolute values. Another interesting feature is that at the peak of a resonance $(E = E_r)\sigma$ can have the order of magnitude $\pi\lambda^2$. At very low energies, since $\lambda^2 \propto 1/p^2 \propto 1/E$, this can be much greater than πR^2 where R is the nuclear radius, and very large values for σ are to be expected. This is illustrated in Fig. 5.12 for the capture of low energy neutrons by ^{113}Cd.

Experimentally it is found that the Γ_i have values in the range from $\simeq 0.1$ eV in the case of low energy incident neutrons through to $\simeq 10^5$ eV for very light target nuclei. These uncertainties (ΔE) in the values of resonant energies imply corresponding uncertainties in the lifetimes ($\Delta t \simeq \hbar/\Delta E$) of compound states in the range from 10^{-14} s to 10^{-20} s. (Note that putting $\Delta E = \Gamma_i$ gives, for the decay probability, P, of a compound state $P \simeq 1/\Delta t \simeq \Gamma_i/\hbar$, i.e. proportional to Γ_i as already mentioned.) Of course, as the energy of bombarding particles increases, the compound states become closer in energy so that,

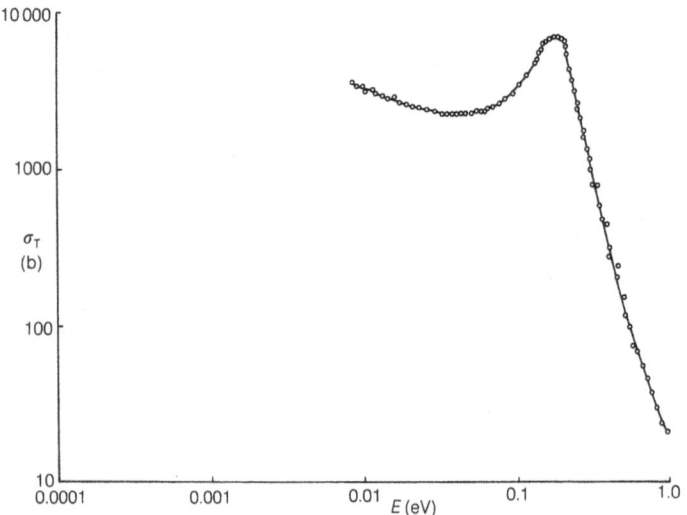

Fig. 5.12 Low energy resonance in ^{113}Cd at $E_r = 0.176$ eV with $\sigma_{res} \simeq 7200$ b and $\Gamma = 0.115$ eV.

because of their widths, they finally overlap and the cross-section is then a smooth function of energy.

Because of the relatively long lifetime of compound states very little information about their mode of formation is carried into the decay channels other than that about conserved quantities such as energy, momentum, angular momentum and charge. As a result angular distributions are rather flat, unlike direct reactions, and tend to be symmetrical about 90°. The main exception is in the case of heavy ion reactions for which very large angular momenta may be involved.

5.9 HEAVY ION REACTIONS

A heavy ion is conventionally defined as one with $A \geqslant 4$ and a heavy ion reaction is one using such an ion as the bombarding particle. The foregoing discussion of different aspects of nuclear reaction processes applies equally to heavy ion reactions. However, this short section dealing specifically with these reactions is included since much current experimental research is concentrated in this area and some interesting phenomena arise.

Heavy ions are in general use with energies up to around 10 MeV per nucleon often from tandem electrostatic accelerators (section 5.3), although energies in the GeV region have also been achieved in high energy synchrotrons. A particular characteristic of these reactions is the high angular momentum involved. For example, a ^{16}O ion with energy 5 MeV per nucleon striking another ^{16}O nucleus peripherally will bring in an angular momentum of $\simeq 40\hbar$.

The Coulomb barrier will, of course, be very large for a high Z ion and so, for low energy ions, Rutherford scattering at low scattering angles modified by diffraction scattering at higher angles (section 5.5) is expected and observed.

At higher energies direct interaction through the nuclear force takes place and this can lead, for example, to inelastic scattering with the excitation of high angular momentum rotational states. Pick-up or stripping of one or more nucleons may also occur.

For even higher energies compound nucleus formation can take place leading to highly excited compound states (hot nuclei) with high angular momenta (e.g. $\simeq 40\hbar$ or more for bombarding ^{16}O ions). For very high angular momenta in high $A (\geqslant 200)$ compound states de-excitation will take place by **fission** in which, under the influence of the strong centrifugal force due to the rotation, the nucleus disintegrates into two large fragments (section 5.10). Otherwise it de-excites by

emitting first neutrons and then photons, eventually reaching a high
rotational level (a **yrast** level – section 4.3) such as that with $J = 60$ in
^{152}Dy. Subsequent de-excitation involves further photon emission as
the nucleus drops down through the levels in rotational bands. Study
of the energies and angular distributions of these photons, as well as
the lifetimes of the nuclear states involved, gives information about
the deformations and moments of inertia of these nuclei, and ratios of
major to minor axes as high as 2:1 (**superdeformed** nuclei) have been
observed.

Another interesting area of investigation is of compound nuclei
formed in heavy ion reactions far from the valley of stability (Fig. 2.3).
Interest has centred, for example, on nuclei in the region $A \simeq 80$ with
$Z \simeq N$ and on **superheavy** nuclei with $A \simeq 300$ which theory predicts
might be stable.

5.10 FISSION

Fission is a process discovered in 1939 by Hahn and Strassmann
in which a heavy nucleus (usually one is thinking of nuclei in the
uranium region) disintegrates into two lighter nuclei together with
two or three neutrons and with a very large release of energy. Before
considering the mechanism of this process we first consider the
energetics.

Referring to the plot of binding energy per nucleon ($B(A, Z)/A$)
against A shown in Fig. 2.2 it can be seen that, whereas in the region A
$\simeq 240$, $B/A \simeq 7.6$ MeV, for $A \simeq 120$, $B/A \simeq 8.5$ MeV. This means that
if an $A \simeq 240$ nucleus divides into two, each nucleon is on average
bound more tightly by around 0.9 MeV and will release that amount
of energy; this corresponds to a total energy release of 240×0.9 MeV
$= 216$ MeV. This is some 10^6 times larger than the energy released in
a chemical process.

In addition, referring to Fig. 2.3 (see also eq. (4.4) *et seq.*), it can be
seen that as A increases the proportion of neutrons in the most stable
nuclei increases. For example, the most stable nucleus with $A = 120$
is $^{120}_{50}$Sn with $N/A = 0.58$ whilst for $A = 240$ the most nearly stable
nucleus is $^{240}_{94}$Pu with $N/A = 0.61$. This effect is due to the increasing
repulsive energy associated with the long-range Coulomb force
between protons and means that, after the fission process, there is an
excess of neutrons in the system. It is some of these neutrons that are
emitted during the fission process (**prompt neutrons**). Others may be
emitted later (**delayed neutrons**) as the fission products adjust

themselves by β^--decay (to reduce further the number of neutrons) into their most stable forms.

The mechanism of the fission process can easily be visualized by thinking of the nucleus as a liquid drop. Diagrams of the successive steps when a nucleus fissions are shown in Fig. 5.13 where the fissioning nucleus starts with an aspherical shape (nuclei with $A \geqslant 225$ have large quadrupole moments – section 4.2.3). There is then further distortion leading to a 'necking' of the nucleus which finally separates (the point of **scission**) into two fragments together with prompt neutrons.

Much light is thrown on the process if a diagram is drawn of the potential energy of the system plotted against its deformation or, past the scission point, the separation between the fission fragments. Such a diagram is given in Fig. 5.14 for a notional nucleus with $A \simeq 240$ and with energy values derived from liquid drop model calculations. As the deformation increases, the potential energy initially increases because work has to be done against the attractive internucleon force (surface tension effect). But increasing distortion and, subsequently, separation between the fragments means that this effect dies away because of the very short range of nuclear forces. However, the long-range Coulomb repulsion between the fragments continues to reduce as they separate, leading eventually to the drop in potential energy of

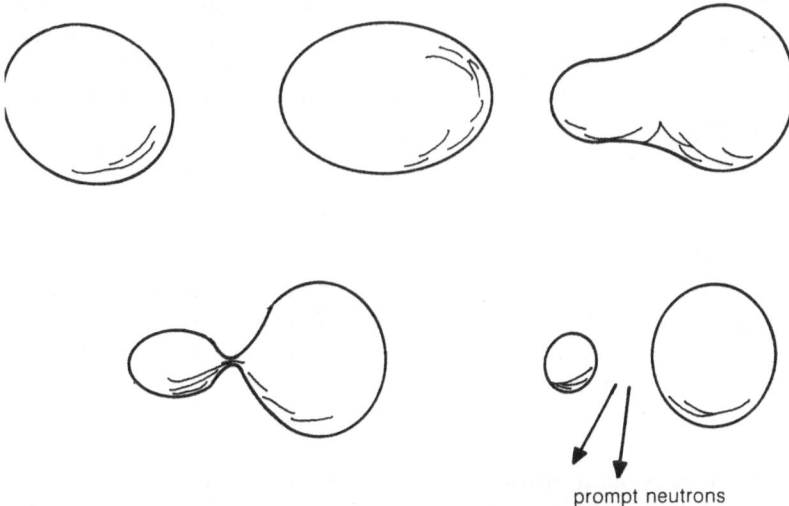

prompt neutrons

Fig. 5.13 Schematic representation of the fission process.

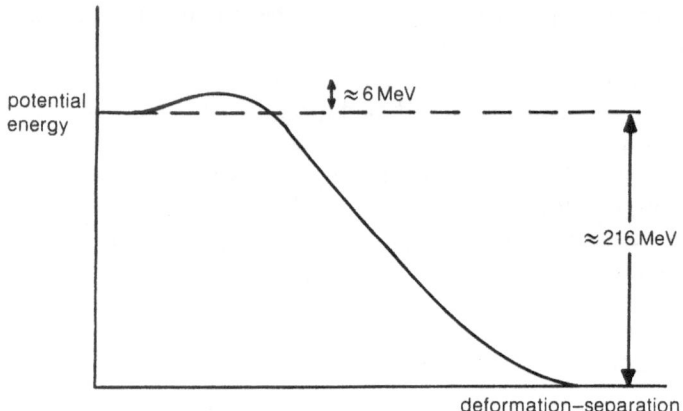

Fig. 5.14 Potential energy in the fission process as a function of deformation-separation.

the order of 216 MeV already mentioned. There is thus a resultant potential barrier of height around 6 MeV which has to be penetrated or surmounted if fission is to take place.

If the barrier is penetrated then **spontaneous fission** takes place. However, the lifetime for this process is very long for nuclei with $A \leqslant 250$ (e.g. for $^{238}_{92}\text{U}$ $\tau \simeq 10^{16}$ years, whilst for $^{256}_{100}\text{Fm}$ $\tau \simeq 10^{-4}$ years) unless the nucleus is in an excited state. Much more interesting is the process of **induced fission** which can be brought about by the capture of a neutron. There are two possibilities as illustrated in the following processes which can occur in natural uranium (99.3% ^{238}U + 0.7% ^{235}U).

For the process

$$^{238}\text{U} + \text{n} \rightarrow {}^{239}\text{U*}$$

the compound nucleus will have excitation energy of $\simeq 5$ MeV for zero-energy neutrons, which is 1 MeV below the potential barrier. Therefore fast neutrons (**fast neutron fission**) with energy around 1 MeV are needed if the barrier is to be surmounted and fission to take place. On the contrary, for the process

$$^{235}\text{U} + \text{n} \rightarrow {}^{236}\text{U*}$$

zero-energy neutrons lead to an excitation of the compound nucleus of $\simeq 6.4$ MeV. This is enough to surmount the barrier and hence fission can be induced by slow neutrons (**thermal neutron fission**).

Higher excitation energy is achieved in this case because ^{236}U is an even–even nucleus in which neutrons are more tightly bound so that, conversely, capture of a neutron leads to a larger release of energy.

Although discussion has been in terms of a nucleus fissioning into equal parts, this is not the situation experimentally. For uranium nuclei the fission products are, for example, varied in mass but centre around $A \simeq 95$ and $A \simeq 135$. Typical fission products for uranium are

$$_{92}U \rightarrow {}_{52}Te + {}_{40}Zr$$
$$\rightarrow {}_{56}Ba + {}_{36}Kr$$
etc.

The situation is clearly more complicated than described by the simple liquid drop model. This is also true for the form of the fission barrier (Fig. 5.14) and more detailed calculations indicate that the barrier has a double hump. These are refinements, however, and the foregoing description contains the essentials for understanding the fission process.

5.10.1 Chain Reactions and Nuclear Reactors

A chain reaction makes use of the fact that the two or three prompt neutrons emitted in a fission process could themselves be used to induce further fission and so on and so on. Of course, considering a block of uranium (say), these neutrons escape or may be lost to fission processes in some other way and a chain reaction will only be sustained if at least $N = 1$ neutron per fission induces another fission. In this situation the assembly is said to be **critical**. For $N < 1$ it is **subcritical** and for $N > 1$ it is **supercritical** and an explosion results.

Chain reactions are the basic processes on which nuclear reactors depend and, of course, nuclear weapons. In the latter, two subcritical masses are rapidly brought together so that the resultant mass is supercritical. There are basically two types of reactor – **thermal** and **fast**. In the former, use is made of the small percentage (0.7%) of ^{235}U in natural uranium and the fact that the cross-section for thermal fission (at neutron energies $\simeq 0.025$ eV) is very large ($\simeq 550$ b) owing to the $\pi\lambda^2$ factor discussed in section 5.8. Fission neutrons, however, have energies around 1 MeV and so must be slowed down quickly to the low thermal energies needed. Energy is most effectively lost by scattering from a body (known as the **moderator**) whose mass is close to that of a neutron, e.g. hydrogen, deuterium, etc. Water, which has a high hydrogen content, has the disadvantage that it captures neutrons through the reaction (radiative capture)

$$n + p \rightarrow p + \gamma$$

and can only be used when the nuclear fuel is highly enriched with ^{235}U. Other moderators in common use are heavy water (D_2O) and graphite (carbon).

It is essential to control the rate at which fission proceeds and this is accomplished using **control rods**. These are usually rods of boron or cadmium which have very high cross-sections for thermal neutron capture (e.g. as shown in Fig. 5.14) and which can be inserted into or withdrawn from the reactor so as to control the number, N, of fissioning neutrons. The heat output of the reactor, deriving from the kinetic energy of the fission fragments, is removed via a circulating coolant which can be carbon dioxide gas or pressurized water.

In a thermal reactor the ^{238}U does not fission but it does capture neutrons with emission of a γ-ray and, following a sequence of β^--decay processes, fissionable ^{239}Pu is produced:

$$^{238}_{92}U + n \xrightarrow[\gamma]{} {}^{239}_{92}U \xrightarrow[\beta]{} {}^{239}_{93}Np \xrightarrow[\beta]{} {}^{239}_{94}Pu$$

^{239}Pu is used together with ^{238}U in **fast reactors**. Here no moderator is used so that the neutrons have energies $\simeq 1$ MeV and the reactor core is much more compact. The rate of heat production is then so high in relation to the size of the core that liquid sodium has to be used as a coolant. Further, neutron capture by ^{238}U as well as producing fission also leads to the production of plutonium, as already described, which can more than replace that removed by fission. Such reactors are therefore referred to as **breeder reactors**. Thermal reactors will not function in this way because at thermal energies too many neutrons are absorbed by ^{239}Pu through radiative capture.

Thermal reactors are a prolific source of thermal neutrons which are used in many experimental studies and also of antineutrinos (section 6.5) resulting from the β^--decay of fission products.

5.11 FUSION AND STELLAR ENERGY

Fusion refers to processes in which two nuclei coalesce and energy is released. Such reactions are primarily responsible for the generation of energy in stars and the massive energy release in a 'hydrogen bomb'. Much effort is being expended in trying to develop efficient controlled nuclear fusion for the provision of energy on earth.

Referring to the plot of B/A against A (Fig. 2.2), the energy released in fission results from a move to more tightly bound nuclei from right to left (heavy nuclei fission into medium weight nuclei). Fusion, on the contrary, involves a move from left to right (fusion of very light nuclei to form somewhat heavier nuclei) and the very tightly bound ^{4}He nucleus plays a key role. The fusion reactions responsible for 'hydrogen burning' in stars (the **hydrogen cycle**) are

$$p + p \rightarrow d + e^+ + \nu \qquad + 0.42\,\text{MeV}$$
$$d + p \rightarrow {}^3\text{He} + \gamma \qquad + 5.49\,\text{MeV}$$
$${}^3\text{He} + {}^3\text{He} \rightarrow {}^4\text{He} + p + p \qquad + 12.86\,\text{MeV}$$

where e^+ signifies a positron, which has the same mass as an electron but opposite charge, and ν signifies a neutrino with mass $\simeq 0$ (section 6.5.4).

Each reaction uses nuclei produced in the previous reaction and the overall outcome of the cycle is that four protons are converted into a helium nucleus (^{4}He) together with two positrons ($2e^+$) and two neutrinos (2ν). The net energy release is, therefore,

$$E = [4m_p - M(^4\text{He}) - 2m_e]c^2 = 24.7\,\text{MeV} \qquad (5.27)$$

where the neutrino mass has been taken to be zero.

The first reaction involves the β-decay process (discussed in Chapter 6) since the reaction $p + p \rightarrow {}^2\text{He} + \gamma$ is forbidden, ^{2}He not being a stable nucleus. It is a **weak** reaction and consequently has a very small cross-section. In addition, the protons have to overcome a Coulomb potential barrier of the order of 1 MeV (section 5.5) in order to interact, an even higher barrier having to be overcome in the third reaction. The temperature in the interior of a star such as the sun is $\simeq 10^7$ K so that the mean kinetic energy of a proton is $3kT/2 \simeq 10^3$ eV where k is the Boltzmann constant. This is well below the barrier height. However, the Boltzmann distribution of energies coupled with quantum barrier penetration effects mean that the reactions can nevertheless proceed at stellar temperatures but the rate is very slow. It is the sheer size of a star that leads to its high energy output. For reactions of this kind to proceed in an, inevitably, relatively infinitesimally smaller fusion reactor on earth requires much higher temperatures, at least $\simeq 10^8$ K. The problem facing researchers in this area is to find ways of producing and containing plasma at temperatures of this kind.

In a star other fusion reactions take place, leading to the formation of different elements. Thus, two ^{4}He nuclei can combine to form ^{8}Be which, in turn, can capture another ^{4}He nucleus to form ^{12}C. The latter nucleus plays a key energy production role in some stars through another set of fusion reactions collectively referred to as the **carbon cycle**:

$$^{12}C + p \rightarrow \,^{13}N + \gamma \qquad + 1.94 \, MeV$$

$$^{13}N \rightarrow \,^{13}C + e^+ \qquad + \nu + 1.20 \, MeV$$

$$^{13}C + p \rightarrow \,^{14}N + \gamma \qquad + 7.55 \, MeV$$

$$^{14}N + p \rightarrow \,^{15}O + \gamma \qquad + 7.29 \, MeV$$

$$^{15}O \rightarrow \,^{15}N + e^+ \qquad + \nu + 1.73 \, MeV$$

$$^{15}N + p \rightarrow \,^{12}C + \,^4He \quad + 4.96 \, MeV$$

As with the hydrogen cycle the carbon cycle effectively leads to the transmutation of four protons into ^4He and leads essentially to the same release of energy.

There are many other fusion reactions that can take place in stars, leading to the formation of even heavier nuclei and also the production of neutrons. A series of neutrons captures then ultimately leads to the formation of the heaviest nuclei.

5.12 COMMENT

Clearly nuclear reactions are extremely complicated and varied, and inevitably the treatment given in this chapter has in general been qualitative. Indeed, in some areas it has been no more than superficial, but it is hoped that it has nevertheless given some impression of the ideas and activities relating to this vitally important area for the study of nuclear phenomena. To go more deeply into these matters requires a significant 'change of gear' and much specialist study.

constant. The underlying assumption in this law is that the probability of decay is governed by the laws of chance and is independent of the past history of the nucleus. For an assembly of N radioactive nuclei, during time dt the number decaying will be $N\,dP$ or, using eq. (6.1)

$$-dN = \lambda N\,dt \qquad (6.2)$$

where $-dN$ is the decrease in the number of nuclei present. Integrating eq. (6.2) then gives for the number, $N(t)$, of nuclei present at time t

$$N(t) = N(0)e^{-\lambda t} \qquad (6.3)$$

where $N(0)$ is the number present at $t = 0$.

The **activity** A of a radioactive substance at time t is defined as the number of decays per unit time and is given by

$$A(t) = \left|\frac{dN}{dt}\right| = \lambda N(t) = \lambda N(0)e^{-\lambda t} \qquad (6.4)$$

using eqs (6.2) and (6.3). Thus the activity of a radioactive source should die away exponentially. Such behaviour was originally observed by Rutherford and Soddy and is now a well-confirmed feature of radioactivity. Note that the conventional unit of activity is the **becquerel** (Bq) which corresponds to one decay per second. (Previously the standard unit of radioactivity in use was the curie (Ci) which was the amount of a radioactive substance for which there were 3.7×10^{10} decays per second.)

It is now straightforward to define the **mean life** and the **half-life** (or **period**) of a radioactive substance. The mean life, τ, is the name given to the average lifetime of a nucleus and is readily calculated. From eq. (6.4) it follows that the number of nuclei decaying between times t and $t + dt$, i.e. with a lifetime between t and $t + dt$, is

$$|dN| = \lambda N(0)e^{-\lambda t}\,dt \qquad (6.5)$$

The mean life is then simply

$$\tau = \frac{1}{N_0} \int_{N(0)}^{0} t\,|dN|$$

$$= \int_{0}^{\infty} \lambda t e^{-\lambda t}\,dt$$

$$= \frac{1}{\lambda} \qquad (6.6)$$

6

Alpha, beta and gamma decay

As discussed in Chapter 1, the radioactivity of heavy atoms, embodied in the three processes of α-, β- and γ-decay, were early indicators of nuclear activity. With the production of many species of nuclei in their ground or excited states through nuclear reactions such decays have been studied throughout the periodic table. These studies, particularly of γ-decay, give valuable information about the structure of nuclear states as well as about the interactions responsible for them.

All three processes involve spontaneous emission from an unstable nucleus. α-decay is the emission of a ^{4}He nucleus (α-particle); β-decay is the emission of an electron (β$^-$-decay) or a positron (β$^+$-decay) accompanied by a 'neutrino' (section 6.5.4) and γ-decay the emission of a photon. Although these processes are treated together in this chapter it must be emphasized that the underlying interactions are quite different although each interaction may derive from some grand unified scheme (Chapter 9). Decay processes of this kind are all, however, characterized by the lifetime of the decaying nucleus and, before discussing the three processes in detail, we must first formalize the meaning of 'lifetime'.

6.1 NUCLEAR LIFETIMES AND DECAY PROBABILITIES

All thinking about radioactive decay centres around the proposition – the law of radioactive decay – that the probability dP of decay during a time interval dt is simply proportional to dt, i.e.

$$dP = \lambda \, dt \qquad (6.1)$$

where the constant of proportionality, λ, is known as the **decay**

ted in passing that different names were given to nuclei in the
oactive series from those used above. For example in the ^{232}Th
s just discussed ^{216}Po was referred to as Th A, ^{212}Pb as Th B,...
^{208}Pb as Th D. Similarly in the series stemming from ^{238}U, ^{218}Po
referred to as Ra A, ^{214}Pb as Ra B, ... and ^{206}Pb as Ra G. This
inology will not be used further in this book.

e study of naturally occurring radioactive nuclei has played an
ortant role in the past in achieving an understanding of radio-
e processes. However, with the advent of nuclear accelerators
ecame possible to produce many artificial radioactive nuclei
ugh nuclear reactions and present-day studies of α-, β- and
cay are largely concentrated on these.

6.3 ALPHA DECAY

cay differs fundamentally from β- and γ-decay insofar as the latter
esses involve the creation of particles not permanently present in
ucleus (electrons, neutrinos, photons) whilst the former is simply
arrangement of the nucleons into a state of lower energy. This
rangement involves the emission of a helium nucleus (an α-
icle) and is completely akin to the decay of a compound nucleus
scussed in section 5.7. The reason that a helium nucleus is emitted
er than, say, a proton or a deuteron is simply that the helium
eus is very tightly bound ($B = 28.3$ MeV – Fig. 2.2) and thus more
gy, Q, is released. Indeed energy would have to be provided (Q
tive) in order for the emission of other nucleon groupings to take
e. Take, for example, the α-decay of ^{228}Th which occurs in the
oactive series discussed in the previous section and consider the
ibility that it might decay by emission of a proton. For α-decay
have

$$^{228}_{88}\text{Th} \rightarrow {}^{224}_{88}\text{Ra} + \alpha \quad (Q \simeq +5.5 \text{ MeV})$$

st for proton emission

$$^{228}_{88}\text{Th} \rightarrow {}^{227}_{87}\text{Ac} + \text{p} \quad (Q \simeq -6.4 \text{ MeV})$$

feature of α-decay that can readily be understood is what was
inally referred to as 'fine structure'. Here α-particles with several
rete energies are sometimes observed in a transition between two
lei. This can be attributed to α-decay taking place to excited states
e daughter nucleus. These usually de-excite by emitting γ-rays

Equation (6.3) can therefore be rewritten

$$N(t) = N(0)e^{-t/\tau} \tag{6.7}$$

The half-life $t_{1/2}$ is defined as the time over which half the nuclei in a
sample will have decayed. It is obtained directly from eq. (6.7) by
writing $N(t) = N(0)/2$ and sòving for t, i.e.

$$\frac{N(0)}{2} = N(0)e^{-t/\tau}$$

giving

$$t = t_{1/2} = (\ln 2)\tau = 0.693\tau \tag{6.8}$$

Care must be taken using lifetimes that it is clear whether half or mean
lifetimes are under discussion.

Observed values of τ vary tremendously ranging from as long as
$\tau \simeq 10^{10}$ years for some naturally occurring α-emitters, through the
range from $\tau \simeq 10^6$ years to $\tau \simeq 10^{-3}$ s for β-emitters down to as short
as $\tau \simeq 10^{-15}$ s for some γ-emitters. Even shorter lifetimes ($\simeq 10^{-20}$ s)
have already been noted (section 5.8) for the disintegration of
compound nuclei into other nuclei–particles and, at the other
extreme, for what are known as **double β-emitters** (section 6.5.4), lifetimes of the order of 10^{21} years or more have been measured.

Frequently in discussing decay probabilities of nuclei and, indeed,
of elementary particles, the **width**, Γ, of a state is used rather than its
lifetime. The concept of the width of a state has already been discussed
in connection with compound nucleus reactions (section 5.8). It
relates to the uncertainty in the time for which a state will live. This is
given by $\Delta t \simeq \tau$ and, by Heisenberg's uncertainty relation, implies a
corresponding uncertainty in the energy of the state: $\Delta E \simeq \hbar/\Delta t \simeq \hbar/\tau$.
This uncertainty in the energy is referred to as the width and so we
have

$$\Gamma \simeq \frac{\hbar}{\tau} \tag{6.9}$$

At two extremes we have for $\tau = \infty$ (a stable nucleus) $\Gamma = 0$, whilst, for
$\tau \simeq 10^{-20}$ s, $\Gamma \simeq 0.1$ MeV. The important point to note is that the
energy of an unstable nuclear state does not have a sharp value and,
although Γ may be small compared with the spacing between energy
levels for a relatively long-lived state, it cannot be neglected for a
short-lived state.

6.1.1 The Measurement of Nuclear Lifetimes

The obvious approach is to measure the activity A of a decaying nucleus as a function of time and to match this to the exponential expression given in eq. (6.4), so determining λ. The simplest way of doing this is to take the logarithm of eq. (6.4), i.e.

$$\ln A(t) = \ln \lambda N(0) - \lambda t \qquad (6.10)$$

and to plot $\ln A(t)$ against t. The slope of the resultant straight line is then equal to $-\lambda = -1/\tau$. This is fairly straightforward for lifetimes in the range from minutes to years. However, as the lifetime shortens, increasingly complicated experimental techniques have been devised to measure the change in activity. These all involve identification of the instant the decaying nucleus is formed, for example in the target of a nuclear reaction experiment, and measurement of the level of activity at some time which may be as little as 10^{-15} s later! For γ-decay such techniques often use beams of excited nuclei produced, usually as ions, following recoil from a target.

For very long lifetimes we have to first order in $\lambda = 1/\tau$ from eq. (6.4)

$$A(t) = \frac{N(0)}{\tau}$$

$$= \frac{N(t)}{\tau} \qquad (6.11)$$

to the same order of approximation, so that from a knowledge of the amount of radioactive substance present and its activity, τ can be deduced at once. In these cases the activity is frequently obtained indirectly from measurements of the quantity of 'daughter' nuclei present rather than measurements on the radiation itself.

At the other extreme, lifetimes in the range from 10^{-15} s through to 10^{-20} s or even shorter are obtained indirectly by measuring the energy distribution of the decay products which gives information about the width, Γ, of the decaying state. This, in turn, enables τ to be estimated using $\tau \simeq \hbar/\Gamma$ (eq. (6.9)).

6.2 NATURAL RADIOACTIVITY

By natural radioactivity we mean radioactivity observed in ores etc. occurring naturally in the earth's surface. Since the activity of a radioactive nucleus is determined by its lifetime, τ, it might be thought

that, if it is to occur naturally, τ must have a value the age of the universe ($\simeq 10^{10}$ years). This is ce naturally occurring radioactive nuclei but, as wil lived nuclei also occur naturally as transitory pr decay of long-lived nuclei.

Examples of long-lived, but unstable, ($\tau = 1.4 \times 10^{10}$ years), ^{235}U ($\tau = 7 \times 10^8$ years) and years): These nuclei are all α-emitters and the followed by a series of α- and β-emissions until obtained. Typical of such radioactive series is th from the α-decay of ^{232}Th:

$$^{232}_{90}\text{Th} \xrightarrow{\alpha} {}^{228}_{88}\text{Ra} \xrightarrow{\beta^-} {}^{228}_{89}\text{Ac} \xrightarrow{\beta^-} {}^{228}_{90}\text{Th} \xrightarrow{\alpha} {}^{224}_{88}\text{Ra}$$

$$\xrightarrow{\alpha} {}^{216}_{84}\text{Po} \xrightarrow{\alpha} {}^{212}_{82}\text{Pb} \xrightarrow{\beta^-} {}^{212}_{83}\text{Bi} \begin{cases} \xrightarrow{\beta^-} {}^{212}_{84}\text{Po} \xrightarrow{\alpha} \\ \xrightarrow{\alpha} {}^{208}_{81}\text{Tl} \xrightarrow{\beta^-} \end{cases}$$

It will be noted that for an α-decay A reduce corresponding to the loss of two neutrons and tw the α-particle, whilst for the β^--decays A remain nucleon is emitted) and Z increases by 1 (since charge has been lost to the nucleus). It will also alternative decay paths are available to ^{212}Bi and probability of β^--decay is approximately twice t decay, leading to the branching ratios indicated

The lifetimes of the different decaying nuclei ir tremendously and, compared with the long lif parent nucleus, some will be very short. In th example, the lifetime of ^{212}Po is $\tau = 3 \times 10^{-7}$ s. Th the series is continually functioning in natural su of the long lifetime of the initial parent, the ore wil short-lived nuclei.

All series terminate with a stable nucleus – ² example. ^{208}Pb is, of course, particularly stable si section 4.2.1, it has a 'magic' number of neutrons ((82). For the other long-lived nuclei mentioned ab nuclei are ^{207}Pb and ^{206}Pb in the cases of the initi ^{238}U respectively.

Early studies of radioactivity were carried out understanding of nuclear structure had been ach which are observed, as is to be expected, to have energies equal to the differences in energies of the α-particles.

The major challenge in understanding α-decay is to account for the very wide variation in lifetimes. Referring to the last section we have seen that lifetimes as disparate as 1.4×10^{10} years (^{232}Th) and 10^{-7} s (^{212}Po can arise, differing by a factor of the order 10^{24}! And yet the energies released differ only by a factor of the order 2 ($\simeq 4$ MeV and $\simeq 9$ MeV respectively). This behaviour can in fact be understood very simply in terms of the α-particle penetrating a Coulomb barrier as it escapes from the nucleus.

6.3.1 Simple Theory of Alpha-Decay

A basic understanding of the α-decay process was achieved in 1928–1929 by Gamow and Condon and Gurney who were able to explain the main trend of the experimental data on α-decay which had been established by Geiger and Nuttall in 1911. These latter experimenters had found that ln τ for many decays is approximately proportional to $T^{-1/2}$ where T is the kinetic energy of the emitted α-particle.

The basic assumption of the theory is that a pre-formed α-particle is confined in the nucleus by the potential barrier due to the combined effects of the attractive nuclear force and the Coulomb interaction between the α-particle (charge $2e$) and the daughter nucleus (charge Ze). The probability of decay is then estimated in terms of the frequency with which the α-particle impinges on the barrier whilst inside the nucleus and the probability of penetrating the barrier. This situation is illustrated in Fig. 6.1 where the potential well due to the nuclear interaction has radius R and depth V_0 and has superimposed on it the Coulomb potential energy curve $V_C = 2Ze^2/4\pi\varepsilon_0 r$ for values of $r > R$. Of course this is a crude model of the interactions but it does contain the essential physics.

The highest point of the barrier is at $r = R$ and for natural α-emitters has a value in the approximate range 25–35 MeV, much higher than the kinetic energy, T, of the emitted α-particles – hence the need for barrier penetration. For values of $r < R$ and $r > r_T$ (Fig. 6.1) the α-particle has positive kinetic energy and the wave function is oscillatory. Here r_T is the value of r at which the height of the Coulomb barrier equals T and is therefore given by

$$\frac{2Ze^2}{4\pi\varepsilon_0 r_T} = T \tag{6.12}$$

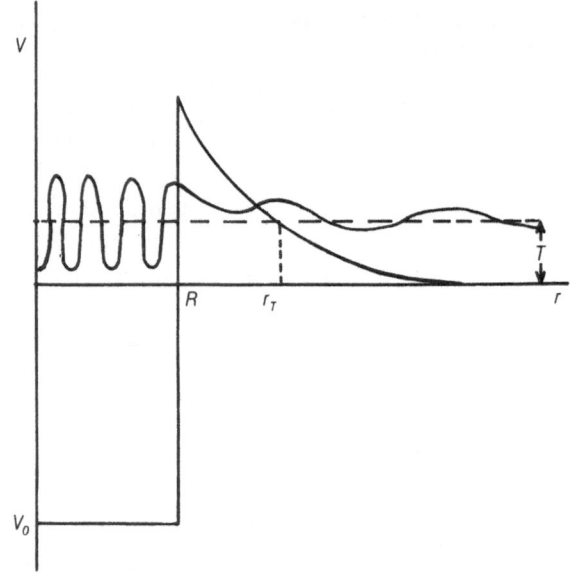

Fig. 6.1 Potential energy diagram for α-decay showing the oscillatory form of the α-particle wave function in the nucleus ($r < R$) and outside the potential barrier ($r > r_T$), and its exponential form within the barrier ($R \leqslant r \leqslant r_T$).

In the classically forbidden region $R \leqslant r \leqslant r_T$ the wave function decays exponentially and the change in its strength over that region of space is a measure of the probability of barrier penetration.

In this region for s-wave α-particles (the simplest situation to consider), the α-particle wave function can be written $\psi(r) = u(r)/r$ where $u(r)$ satisfies the Schrödinger equation

$$-\frac{\hbar^2}{2\mu}\frac{d^2u}{dr^2} + V(r)u = Tu \qquad (6.13)$$

where μ is the reduced mass of the α-particle. Substituting $u = e^{-y}$ leads at once to the equation

$$-\frac{d^2y}{dr^2} + \left(\frac{dy}{dr}\right)^2 = \frac{2\mu}{\hbar^2}(V - T) \qquad (6.14)$$

However, anticipating the exponential decay of u, we expect $y \sim r$, so that $d^2y/dr^2 \simeq 0$, giving

$$\frac{dy}{dr} \simeq \left[\frac{2\mu(V - T)}{\hbar^2}\right]^{1/2}$$

or

$$y \simeq \int \left[\frac{2\mu(V-T)}{\hbar^2} \right]^{1/2} dr \qquad (6.15)$$

This means that the ratio of the values of u at r_T and R is given by

$$\frac{u(r_T)}{U(R)} \simeq e^{-\gamma} \text{ where } \gamma = \frac{1}{\hbar} \int_R^{r_T} [2\mu(V-T)]^{1/2} dr \qquad (6.16)$$

An approximate measure of the barrier penetrability, P, is the ratio of the wave function intensities at r_T and R, i.e. the square of the ratio given in eq. (6.16). Thus

$$P \simeq e^{-2\gamma} \qquad (6.17)$$

To estimate the decay probability λ, P must be multiplied by the frequency, f, with which the α-particle impinges on the barrier whilst inside the nucleus. f has the order of magnitude $v_i/2R$ where v_i is the speed of the α-particle inside the nucleus ($\simeq 10^7 \text{ m s}^{-1}$) and R is the nuclear radius ($\simeq 10^{-14} \text{ m}$ for a heavy nucleus). Thus $f \simeq 10^{21} \text{ s}^{-1}$. The decay probability λ and mean life τ are then given by

$$\lambda \simeq \tau^{-1} \simeq \frac{v_i}{2R} e^{-2\gamma} \qquad (6.18)$$

The final step in obtaining λ is to evaluate the integral in eq. (6.16). This is straightforward, if a little tedious (problem 6.3). However, a good understanding of the remarkable energy dependence of α-decay probabilities can be obtained by making two further approximations in the integration, namely (i) putting $T = 0$ (since as remarked earlier T is much smaller than the height of the Coulomb barrier) and (ii) putting $R = 0$ (since R is generally much smaller than r_T). With these approximations, and using the relationship between r_T and T given in eq. (6.12), the integration gives immediately

$$\gamma = \frac{8Ze^2}{4\pi\varepsilon_0 \hbar v} \qquad (6.19)$$

where v is the speed of the emitted α-particle ($T = \frac{1}{2}\mu v^2$).

Thus, substituting in eq. (6.18) and taking logarithms gives

$$\ln \lambda = -\ln \tau = \ln \left(\frac{v_i}{2R} \right) - \frac{16Ze^2}{4\pi\varepsilon_0 \hbar v} \qquad (6.20)$$

Noting that $v = (2/\mu)^{1/2} T^{1/2}$ means that for a given nucleus eq. (6.20) has the essential form

$$\ln \tau = C + DT^{-1/2} \qquad (6.21)$$

in agreement with the experimental observations of Geiger and Nuttall referred to earlier.

In eq. (6.21) the two constants have the values

$$C = -\ln\left(\frac{v_i}{R}\right) \text{ and } D = \frac{16Ze^2}{4\pi\varepsilon_0\hbar}\left(\frac{\mu}{2}\right)^{1/2} \qquad (6.22)$$

A rather more accurate evaluation of the integral in eq. (6.16) leads to essentially the same result but with the numeral 16 in the numerator of D replaced by 4π (problem 6.3) and this will be used in the following.

The C term in eq. (6.21) has much the same value for all heavy nuclei but the D term, depending on Z and T, can vary significantly. Taking the value of μ for $A \simeq 220$ and expressing T in MeV, eq. (6.21) then conveniently reduces, with reasonable accuracy, to

$$\ln \tau \simeq C + 4ZT^{-1/2} \qquad (6.23)$$

Using this expression it is interesting to compare the mean lives of ^{232}Th (Z(daughter) $= 88$, $T = 4.08$ MeV) and ^{212}Po (Z(daughter) $= 84$, $T = 8.95$ MeV). Using the foregoing values for Z and T gives

$$\ln \tau(\text{Th}) \simeq A + 175$$
$$\ln \tau(\text{Po}) \simeq A + 112$$

Subtracting these two expressions then gives

$$\ln\left[\frac{\tau(\text{Th})}{\tau(\text{Po})}\right] \simeq 63 \quad \text{or} \quad \frac{\tau(\text{Th})}{\tau(\text{Po})} \simeq e^{63} \simeq 2.3 \times 10^{27}$$

Experimentally this ratio, as has already been observed, is $\simeq 10^{24}$ but, considering the very approximate nature of the theory presented, the agreement is reasonable. Clearly the theory can be, and has been, significantly refined and there is general agreement with experiment. But the basic physics remains the same and the large variation in α-decay lifetimes can be attributed to the extreme sensitivity of barrier penetrability to the energy of the α-particle.

6.4 GAMMA DECAY

We have seen in Chapter 4 that nuclei have many excited states characterized by energy (E), spin (J) and parity $(P = \pm 1)$. If a nucleus

is left in one of these states following a nuclear reaction, radioactive α-
or β-decay or some other process, it usually de-excites by dropping to
a lower level and emitting a photon. This process is referred to as
gamma decay (γ-decay) and is completely analogous to the emission
of radiation by an atom. However, because energy level separations
are generally much smaller in atoms than in nuclei (typically
electronvolts compared with megaelectronvolts), it follows that
the corresponding wavelengths are significantly different (λ_{atomic}
$\sim 10^{-7}$m compared with $\lambda_{\text{nuclear}} \sim 10^{-12}$–$10^{-13}$ m). These wavelengths are to be compared with the corresponding sizes (as
measured by their radii) of atoms and nuclei, i.e. $R_{\text{atomic}} \sim 10^{-10}$ m
and $R_{\text{nuclear}} \sim 10^{-14}$ m, so that

$$\left(\frac{R}{\lambda}\right)_{\text{atomic}} \simeq 10^{-3} \quad \text{and} \quad \left(\frac{R}{\lambda}\right)_{\text{nuclear}} \simeq 10^{-1}\text{–}10^{-2} \quad (6.24)$$

As will be seen, the difference in these ratios has profound implications. Thus, in the atomic case, it is only necessary, in general, to
consider electric dipole radiation whilst, in the nuclear case, many
types of multipole radiation play an important role. We now go on to
consider the nature of such radiation.

6.4.1 Multipole Radiation

Electric dipole radiation (denoted by E1 in nuclear physics) in atoms
and nuclei is the quantum equivalent of the radiation produced
classically by an oscillating electric dipole. For a particle of charge e
at coordinate r the electric dipole moment is er and quantum
mechanically the amplitude for an electric dipole transition between
two states u_i and u_f is proportional to the electric dipole matrix
element M_{if} given by

$$M_{if} = \int u_f{}^* er u_i \, d\tau \qquad (6.25)$$

where $d\tau$ is an element of volume. The above expression applies, for
example, to the transition of a single proton between two states and,
for a multinucleon wave function, er is simply replaced by a sum of
such terms over all protons in the nucleus.

The transition can be represented diagrammatically as in Fig. 6.2a.
Similarly, photon absorption can be represented as in Fig. 6.2b. The
overall strength of the transition is conditioned by the value of e
(written at the vertex of the diagram) and which, in this context, is

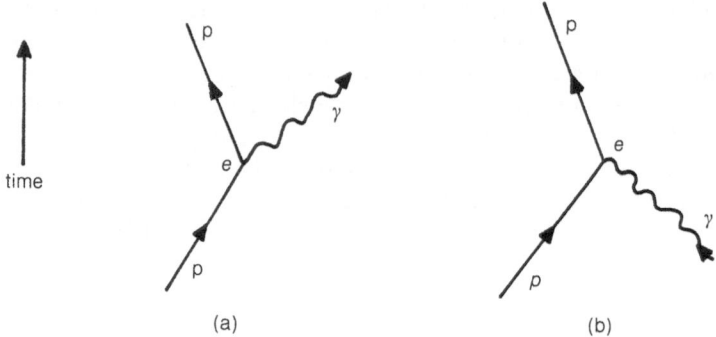

Fig. 6.2 Diagram representing (a) emission and (b) absorption of a photon by a proton.

referred to as the **electromagnetic coupling constant**. Diagrams such as this – generally known as Feynman diagrams – are used to symbolize all forms of electromagnetic transition and, as will be seen, many other forms of elementary particle process.

Returning to the electric dipole transition, the transition probability, T (described in section 6.1 where it is denoted by λ), is proportional to $|M_{if}|^2$ and the form of the proportionality factor can be determined from simple dimensional considerations. Dimensionally, this factor would be expected to depend only on $\lambdabar$, c and the energy E_γ of the emitted photon. Noting that λ has the dimensions (time)$^{-1}$ and that M_{if} has the dimensions (charge × length) it is straightforward (problem 6.3) to show that

$$T(E1) \propto \frac{1}{4\pi\varepsilon_0\lambdabar}\left(\frac{E_\gamma}{\lambdabar c}\right)^3 |M_{if}|^2 \qquad (6.26)$$

with e expressed in coulombs. This result is in agreement in form with that from a full calculation. The magnitude of the dipole matrix element will be $\simeq eR$ where R is the nuclear radius so that

$$T(E1) \propto \frac{e^2}{4\pi\varepsilon_0\lambdabar}\left(\frac{E_\gamma}{\lambdabar c}\right)^3 R^2 \propto \frac{R^2}{\lambda^3} \qquad (6.27)$$

using $E_\gamma = h\nu = hc/\lambda$.

The next highest form of electric multipole radiation is that equivalent to the oscillation of an electric quadrupole moment (E2 radiation) and here the amplitude for a transition is determined by the matrix element between two states of the electric quadrupole operator. Nuclear quadrupole moments have been discussed in section 2.7 where it is noted that they have the order of magnitude eR^2 so that, on dimensional grounds, for an E2 transition,

$$T(E2) \propto \frac{e^2}{4\pi\varepsilon_0\hbar} \left(\frac{E_\gamma}{\hbar c}\right)^5 R^4 \propto \frac{R^4}{\lambda^5} \qquad (6.28)$$

Exactly analogous arguments can be applied for higher electric multipoles (octupole, E3; hexadecupole, E4; etc.) and in general for an EL transition

$$T(EL) \propto \frac{e^2}{4\pi\varepsilon_0\hbar} \left(\frac{E_\gamma}{\hbar c}\right)^{2L+1} R^{2L} \propto \frac{1}{\lambda}\left(\frac{R}{\lambda}\right)^{2L} \qquad (6.29)$$

Using this expression it follows that the ratio of the transition probabilities for neighbouring multipoles is of the order

$$\frac{T(EL+1)}{T(EL)} \propto E_\gamma{}^2 R^2 \propto \left(\frac{R}{\lambda}\right)^2 \qquad (6.30)$$

Referring to the introductory paragraph it can be seen that this ratio is of the order 10^{-2}–10^{-4} for nuclei compared with 10^{-6} for an atom. Hence the relative unimportance of higher multipoles in atomic phenomena.

In addition to electric there can also be magnetic multipole radiation corresponding to the oscillation of a magnetic dipole (M1), magnetic quadrupole (M2) and so on. Quantum mechanically the transition amplitudes will now depend on matrix elements of magnetic operators which will involve electric currents rather than charge. Writing $j \simeq ev$, where v is a typical nucleon speed in the nucleus, the order of magnitude of these matrix elements can then be estimated by replacing e by ev/c in the above expressions for transition probabilities. The uncertainty relation $\Delta p\,\Delta x \simeq \hbar$ can be used to estimate v using $\Delta p \simeq m_p v$ and $\Delta x \simeq R$ which gives $v/c \simeq \hbar/m_p Rc$. There is, however, an additional large contribution from the intrinsic magnetic moments μ_p and μ_n of proton and neutron (section 2.7.1) which introduce a multiplicative factor of the order 10 (note that in units of the nuclear magneton $\mu_p^2 \simeq 8$). Thus comparing transition probabilities for an ML and an EL transition we expect

$$T(ML) \simeq 10\left(\frac{\hbar}{m_p Rc}\right)^2 T(EL) \qquad (6.31)$$

Equations (6.29) and (6.31) give a good guide to the energy dependence and relative importance of different forms of multipole radiation, but detailed calculation is necessary to obtain absolute values. Weisskopf, for example, in the 1950s obtained the above results for single-particle transitions but with an additional multiplicative constant, $18(L+1)/L(L+3)^2[(2L+1)!!]^2$ in eq. (6.29). Using this constant, plots of the half-life, $t_{1/2} = (\ln 2)/T$ (eqs (6.6)–(6.8)), against energy, E_γ, can be drawn and these are shown in Fig. 6.3. In due course (section 6.4.3) the extent to which experimental values of $t_{1/2}$ agree with theory will be discussed. But before doing this we need to obtain the selection rules which govern electromagnetic multipole transitions.

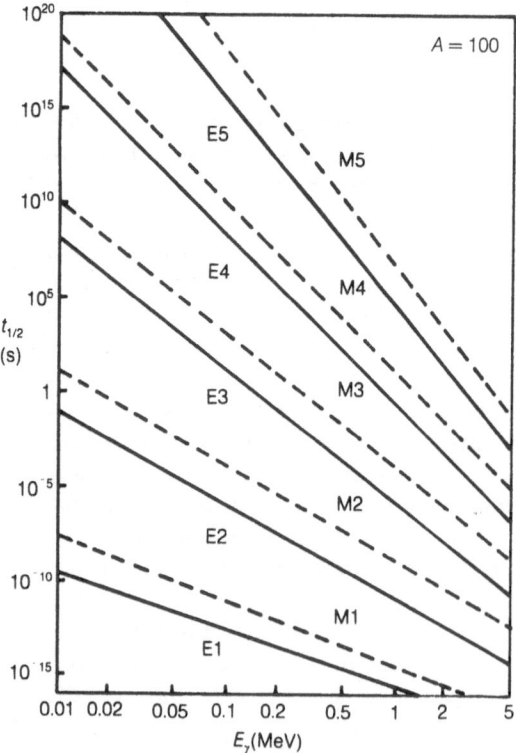

Fig. 6.3 Weisskopf diagram of half-life ($t_{1/2}$) as a function of γ-ray energy (E_γ) for different multipole transitions for a nucleus with $A = 100$.

Equation (6.3) can therefore be rewritten

$$N(t) = N(0)e^{-t/\tau} \tag{6.7}$$

The half-life $t_{1/2}$ is defined as the time over which half the nuclei in a sample will have decayed. It is obtained directly from eq. (6.7) by writing $N(t) = N(0)/2$ and sòving for t, i.e.

$$\frac{N(0)}{2} = N(0)e^{-t/\tau}$$

giving

$$t = t_{1/2} = (\ln 2)\tau = 0.693\tau \tag{6.8}$$

Care must be taken using lifetimes that it is clear whether half or mean lifetimes are under discussion.

Observed values of τ vary tremendously ranging from as long as $\tau \simeq 10^{10}$ years for some naturally occurring α-emitters, through the range from $\tau \simeq 10^6$ years to $\tau \simeq 10^{-3}$ s for β-emitters down to as short as $\tau \simeq 10^{-15}$ s for some γ-emitters. Even shorter lifetimes ($\simeq 10^{-20}$ s) have already been noted (section 5.8) for the disintegration of compound nuclei into other nuclei–particles and, at the other extreme, for what are known as **double β-emitters** (section 6.5.4), lifetimes of the order of 10^{21} years or more have been measured.

Frequently in discussing decay probabilities of nuclei and, indeed, of elementary particles, the **width**, Γ, of a state is used rather than its lifetime. The concept of the width of a state has already been discussed in connection with compound nucleus reactions (section 5.8). It relates to the uncertainty in the time for which a state will live. This is given by $\Delta t \simeq \tau$ and, by Heisenberg's uncertainty relation, implies a corresponding uncertainty in the energy of the state: $\Delta E \simeq \hbar/\Delta t \simeq \hbar/\tau$. This uncertainty in the energy is referred to as the width and so we have

$$\Gamma \simeq \frac{\hbar}{\tau} \tag{6.9}$$

At two extremes we have for $\tau = \infty$ (a stable nucleus) $\Gamma = 0$, whilst, for $\tau \simeq 10^{-20}$ s, $\Gamma \simeq 0.1$ MeV. The important point to note is that the energy of an unstable nuclear state does not have a sharp value and, although Γ may be small compared with the spacing between energy levels for a relatively long-lived state, it cannot be neglected for a short-lived state.

6.1.1 The Measurement of Nuclear Lifetimes

The obvious approach is to measure the activity A of a decaying nucleus as a function of time and to match this to the exponential expression given in eq. (6.4), so determining λ. The simplest way of doing this is to take the logarithm of eq. (6.4), i.e.

$$\ln A(t) = \ln \lambda N(0) - \lambda t \qquad (6.10)$$

and to plot $\ln A(t)$ against t. The slope of the resultant straight line is then equal to $-\lambda = -1/\tau$. This is fairly straightforward for lifetimes in the range from minutes to years. However, as the lifetime shortens, increasingly complicated experimental techniques have been devised to measure the change in activity. These all involve identification of the instant the decaying nucleus is formed, for example in the target of a nuclear reaction experiment, and measurement of the level of activity at some time which may be as little as 10^{-15} s later! For γ-decay such techniques often use beams of excited nuclei produced, usually as ions, following recoil from a target.

For very long lifetimes we have to first order in $\lambda = 1/\tau$ from eq. (6.4)

$$A(t) = \frac{N(0)}{\tau}$$

$$= \frac{N(t)}{\tau} \qquad (6.11)$$

to the same order of approximation, so that from a knowledge of the amount of radioactive substance present and its activity, τ can be deduced at once. In these cases the activity is frequently obtained indirectly from measurements of the quantity of 'daughter' nuclei present rather than measurements on the radiation itself.

At the other extreme, lifetimes in the range from 10^{-15} s through to 10^{-20} s or even shorter are obtained indirectly by measuring the energy distribution of the decay products which gives information about the width, Γ, of the decaying state. This, in turn, enables τ to be estimated using $\tau \simeq \hbar/\Gamma$ (eq. (6.9)).

6.2 NATURAL RADIOACTIVITY

By natural radioactivity we mean radioactivity observed in ores etc. occurring naturally in the earth's surface. Since the activity of a radioactive nucleus is determined by its lifetime, τ, it might be thought that, if it is to occur naturally, τ must have a value commensurate with the age of the universe ($\simeq 10^{10}$ years). This is certainly true of some naturally occurring radioactive nuclei but, as will be seen, very short-lived nuclei also occur naturally as transitory products following the decay of long-lived nuclei.

Examples of long-lived, but unstable, nuclei are ^{232}Th ($\tau = 1.4 \times 10^{10}$ years), ^{235}U ($\tau = 7 \times 10^8$ years) and ^{238}U ($\tau = 4.5 \times 10^9$ years): These nuclei are all α-emitters and their initial α-decay is followed by a series of α- and β-emissions until a stable nucleus is obtained. Typical of such radioactive series is the following starting from the α-decay of ^{232}Th:

$$^{232}_{90}\text{Th} \xrightarrow{\alpha} {}^{228}_{88}\text{Ra} \xrightarrow{\beta^-} {}^{228}_{89}\text{Ac} \xrightarrow{\beta^-} {}^{228}_{90}\text{Th} \xrightarrow{\alpha} {}^{224}_{88}\text{Ra} \xrightarrow{\alpha} {}^{220}_{86}\text{Rn}$$

$$\xrightarrow{\alpha} {}^{216}_{84}\text{Po} \xrightarrow{\alpha} {}^{212}_{82}\text{Pb} \xrightarrow{\beta^-} {}^{212}_{83}\text{Bi} \begin{cases} \xrightarrow{\beta^-} {}^{212}_{84}\text{Po} \xrightarrow{\alpha} {}^{208}_{82}\text{Pb} & (64\%) \\ \xrightarrow{\alpha} {}^{208}_{81}\text{Tl} \xrightarrow{\beta^-} {}^{208}_{82}\text{Pb} & (36\%) \end{cases}$$

It will be noted that for an α-decay A reduces by 4 and Z by 2 corresponding to the loss of two neutrons and two protons bound in the α-particle, whilst for the β^--decays A remains the same (since no nucleon is emitted) and Z increases by 1 (since one unit of negative charge has been lost to the nucleus). It will also be noted that two alternative decay paths are available to ^{212}Bi and it turns out that the probability of β^--decay is approximately twice the probability of α-decay, leading to the branching ratios indicated.

The lifetimes of the different decaying nuclei in the series can vary tremendously and, compared with the long lifetime of the initial parent nucleus, some will be very short. In the above series, for example, the lifetime of ^{212}Po is $\tau = 3 \times 10^{-7}$ s. This means that, since the series is continually functioning in natural surroundings because of the long lifetime of the initial parent, the ore will contain some very short-lived nuclei.

All series terminate with a stable nucleus – ^{208}Pb in the above example. ^{208}Pb is, of course, particularly stable since, as discussed in section 4.2.1, it has a 'magic' number of neutrons (126) and of protons (82). For the other long-lived nuclei mentioned above the terminating nuclei are ^{207}Pb and ^{206}Pb in the cases of the initial parents ^{235}U and ^{238}U respectively.

Early studies of radioactivity were carried out before our current understanding of nuclear structure had been achieved and it should be noted in passing that different names were given to nuclei in the radioactive series from those used above. For example in the ^{232}Th series just discussed ^{216}Po was referred to as Th A, ^{212}Pb as Th B, ... and ^{208}Pb as Th D. Similarly in the series stemming from ^{238}U, ^{218}Po was referred to as Ra A, ^{214}Pb as Ra B, ... and ^{206}Pb as Ra G. This terminology will not be used further in this book.

The study of naturally occurring radioactive nuclei has played an important role in the past in achieving an understanding of radioactive processes. However, with the advent of nuclear accelerators it became possible to produce many artificial radioactive nuclei through nuclear reactions and present-day studies of α-, β- and γ-decay are largely concentrated on these.

6.3 ALPHA DECAY

α-decay differs fundamentally from β- and γ-decay insofar as the latter processes involve the creation of particles not permanently present in the nucleus (electrons, neutrinos, photons) whilst the former is simply a rearrangement of the nucleons into a state of lower energy. This rearrangement involves the emission of a helium nucleus (an α-particle) and is completely akin to the decay of a compound nucleus as discussed in section 5.7. The reason that a helium nucleus is emitted rather than, say, a proton or a deuteron is simply that the helium nucleus is very tightly bound ($B = 28.3$ MeV – Fig. 2.2) and thus more energy, Q, is released. Indeed energy would have to be provided (Q negative) in order for the emission of other nucleon groupings to take place. Take, for example, the α-decay of ^{228}Th which occurs in the radioactive series discussed in the previous section and consider the possibility that it might decay by emission of a proton. For α-decay we have

$$^{228}_{88}\text{Th} \rightarrow {}^{224}_{88}\text{Ra} + \alpha \quad (Q \simeq +5.5\,\text{MeV})$$

whilst for proton emission

$$^{228}_{88}\text{Th} \rightarrow {}^{227}_{87}\text{Ac} + \text{p} \quad (Q \simeq -6.4\,\text{MeV})$$

A feature of α-decay that can readily be understood is what was originally referred to as 'fine structure'. Here α-particles with several discrete energies are sometimes observed in a transition between two nuclei. This can be attributed to α-decay taking place to excited states in the daughter nucleus. These usually de-excite by emitting γ-rays

6.4.2 Multipole Selection Rules

For a given form of multiple radiation (E1, E2,..., M1, M2,...) to take place the matrix element of the corresponding operator must be non-vanishing. This depends on the nature of the wave functions and the symmetry of the operator and it is with the latter that we now concern ourselves. Take, for example, the E1 operator, *er*. Being a vector, this operator has three spatial components and it also has odd parity (under reflection of axes $r \rightarrow -r$). It is to be expected, therefore, that the emitted multipole radiation deriving from this operator reflects these symmetry properties and will be characterized by an angular momentum quantum number $L = 1$ (with three components $M_L = 0, \pm 1$) and odd parity. (To reinforce this assertion, note that the z component of *er* is $ez = er \cos \theta \sim er Y_{10} (\cos \theta)$ which is an angular momentum eigenfunction with $L = 1$ and odd parity. Similarly the x and y components can be written in terms of $Y_{1 \pm 1}$.) In the usual notation a 1^- photon is emitted. But angular momentum and parity must be conserved in the transition so that for an electric dipole transition to take place the changes in nuclear spin, ΔJ, and parity, ΔP, must satisfy

$$\Delta J = 0, \ \pm 1 \ (0 \nrightarrow 0), \text{ to conserve angular momentum}$$

and

$$\Delta P = \text{'yes', to conserve parity}$$

where the inhibition $0 \nrightarrow 0$ arises because angular momentum conservation does not allow one unit of angular momentum to be carried away in a transition between two states with $J = 0$; the triangular rule for angular momentum vectors cannot be satisfied. The above are the selection rules for an E1 transition.

For an M1 transition we again have a vector operator, μ, with three components but which, behaving like an angular momentum ($\sim r \times p$), has even parity. Thus the M1 selection rules are

$$\Delta J = 0, \pm 1 \ (0 \nrightarrow 0)$$

and

$$\Delta P = \text{'no'}$$

For an E2 transition, it will be noted that the spatial part of the electric quadrupole operator given in section 2.7 is $3z^2 - r^2 \sim r^2 Y_{20}$ and has even parity. This operator obviously corresponds to $L = 2$, $P = $ even, so that a 2^+ photon is emitted. The corresponding selection

Table 6.1 Multipole selection rules

Multi-pole	L^P	ΔJ	ΔP	Multi-pole	L^P	ΔJ	ΔP
E1	1^-	$0, \pm 1$	yes	M1	1^+	$0, \pm 1$	no
E2	2^+	$0, \pm 1, \pm 2$	no	M2	2^-	$0, \pm 1, \pm 2$	yes
E3	3^-	$0, \pm 1, \pm 2,$ ± 3	yes	M3	3^+	$0, \pm 1, \pm 2,$ ± 3	no
E4	4^+	$0, \pm 1, \pm 2,$ $\pm 3, \pm 4$	no	M4	4^-	$0, \pm 1, \pm 2,$ $\pm 3, \pm 4$	yes
etc.				etc.			

rules are therefore

$$\Delta J = 0, \pm 1, \pm 2 \, (0 \nrightarrow 0, 0 \nrightarrow 1, 1 \nrightarrow 0, 1/2 \nrightarrow 1/2)$$
$$\Delta P = \text{'no'}$$

Similar arguments can be developed for higher multipoles and the results, excluding the detailed inhibitions given in the above examples, are listed in Table 6.1.

Usually, in considering a transition between two states, because of the decreasing probability of a transition with increasing multipolarity, the lowest possible multipole allowed by angular momentum and parity conservation dominates. However, as will be discussed in the next section, E2 transitions, in particular, are frequently much stronger than predicted on a single-particle model and, because

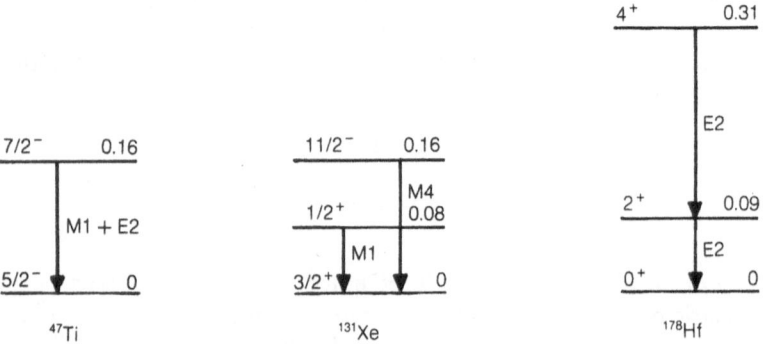

Fig. 6.4 Typical γ transitions.

of this, there are many transitions in which M1 and E2 multipoles both contribute. In Fig. 6.4, examples of the transitions to be expected in various cases are illustrated.

Note that since the electromagnetic field is a vector field (deriving from the three-component vector potential A), the photon can be regarded as having intrinsic spin 1 (three components). The different values of L in Table 6.1 are then to be seen as combinations of this intrinsic spin and the photon's orbital angular momentum relative to the emitting nucleus. Further, the intrinsic parity of the photon must be that of the electromagnetic field A, i.e. odd. Thus the photon is a 1^- particle.

6.4.3 Discussion

We now give brief discussions of a number of topics which arise in considering γ-decay processes.

1. Half-lives – theory and experiment. Figure 6.3 illustrates the wide variation of half-lives to be expected in γ-decay and, by and large, experimental measurements are in fairly general agreement, particularly with the predictions for higher multipoles. Here it is to be noted that if a state can only decay by emitting a high multipole, it will live for a relatively very long time. For example, the half-life of the $11/2^-$ state in ^{131}Xe (Fig. 6.4), which decays to the ground state by an M4 transition, is 11.8 days. States such as this, with half-lives greater than about 1 s, are referred to as **isomeric** states or **isomers**. They tend to occur in regions where low excited states have very high spins and can only decay by high order multipole radiation. Such states occur, for example, in those regions of the periodic table when the shell model $1h_{11/2}$ and $1i_{13/2}$ levels are being filled as mentioned in section 4.2.

   Reference has already been made in this section to the fact that many E2 half-lives are much shorter than predicted by the single-particle model. This is true, for example, for the E2 transitions in ^{180}Hf shown in Fig. 6.4 whose lifetimes are some two orders of magnitude shorter. This can be understood as a collective phenomenon involving contributions to the transition probability from many nucleons. Such effects tend to occur in those regions of the periodic table where the collective model (section 4.3) gives a good description of the energy levels.

2. Internal conversion. Frequently in a γ-decay process (photon energy $= E_\gamma$) electrons are also observed to be emitted with

energies

$$E = E_\gamma - W_i \qquad (6.32)$$

where W_i refers to the binding energy of electrons in the $i =$ K, L, M,... shells of the atom in which the nucleus is embedded. Here it might be thought that an emitted γ-particle gives its energy up to, say, a K electron which then emerges with energy $E_\gamma - W_K$. This would be a sort of internal photoelectric effect. Although this process can happen it is highly improbable and the correct explanation is that energy is transferred directly from the nucleus to an orbiting electron via the electromagnetic interaction (largely the Coulomb term). The process is referred to as **internal conversion** (i.c.) and is an additional mode of nuclear de-excitation. Thus the transition probability, T, for the decay of an excited state has the form

$$T = T_\gamma + T_{i.c.} \qquad (6.33)$$

and an **internal conversion coefficient**, α, can be defined as follows:

$$\alpha = \frac{T_{i.c.}}{T_\gamma} = \frac{N_e}{N_\gamma} \qquad (6.34)$$

where N_e and N_γ are respectively the number of conversion electrons and gamma particles emitted per unit time in the decay. α can also be subdivided and written as

$$\alpha = \alpha_K + \alpha_L + \alpha_M + \ldots \qquad (6.35)$$

where α_K is the K shell conversion coefficient and so on.

It turns out that the values of conversion coefficients are dependent on the multipolarity of a transition, particularly at low energies, and, being a ratio, are largely independent of nuclear details. This means that their measurement is a valuable tool for establishing multipolarities.

There is another form of internal conversion – **pair internal conversion** – which can occur when the decay energy is greater than $2m_e c^2 = 1.02$ MeV, in which the excited nucleus emits an electron–positron pair. A well-known example of this is the decay of the 6.06 MeV (0^+) state of ^{16}O to the 0^+ ground state. Here it will be noted that none of the γ-decay selection rules allows a $0 \rightarrow 0$ transition since photons must carry away at least one unit of angular momentum. For $0^+ \rightarrow 0^+$ transitions the decay can only take place by internal conversion or, if the energy is sufficiently high, by pair internal conversion as in the case just quoted. For $0 \rightarrow 0$ decays involving a change of parity internal conversion is not allowed and the most probable decay mode is the emission of two gamma particles – an E1 and an M1.

3. Angular correlations. The direction in which a γ-ray is emitted in a transition between two nuclear states whose spins and z components are J_i, M_i and J_f, M_f depends on the values of M_i and M_f. In general all magnetic substates of a decaying state are occupied with equal probability and as a result there is no preferred direction of γ-emission.

However, there are ways of arranging for unequal populations of substates. One such method which is widely used is to study the directions of emission of two successive γ-rays (γ_1 and γ_2 – Fig. 6.5a). If γ_1 is always detected in a particular direction, then it will leave the M_f substates unequally populated. Detection of γ_2 in coincidence with γ_1 ensures that it will have been emitted from this unequal population and that its angular distribution relative to the direction of γ_1 will not, in general, be spherically symmetrical. Detailed theoretical study shows that the probability $I(\theta)$ of γ_2 being detected at an angle θ to the direction of γ_1 has the general form

$$I(\theta) = A + B \cos^2 \theta + C \cos^4 \theta + \ldots \tag{6.36}$$

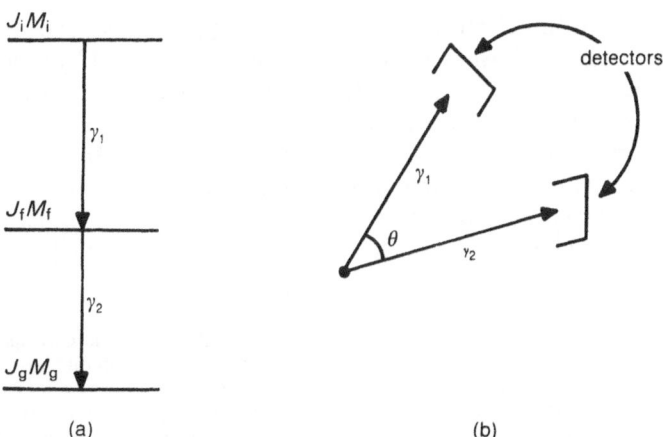

(a)   (b)

Fig. 6.5 An angular correlation experiment in which (a) two successive γ-rays are (b) detected in coincidence as a function of their angular separation θ.

where the values of the coefficients and the maximum power of $\cos \theta$ are dependent on the spins and parities of the nuclear states involved and on the multipolarities. Studies such as this have provided valuable information about the spins and parities of many nuclear levels.

6.5 BETA DECAY

β-decay is a manifestation of what is known as the **weak interaction** (Chapter 9). It is a process in which an unstable nucleus decays into another nucleus and emits either a positron (e^+) and a neutrino (ν_e) or an electron (e^-) and an antineutrino ($\tilde{\nu}_e$), the emitted particles being collectively referred to as **leptons***. These two processes are symbolized as follows:

$$^A_Z X \rightarrow _{Z-1}^{A} Y + e^+ + \nu_e \qquad (\beta^+\text{-decay})$$

$$^A_Z X' \rightarrow _{Z+1}^{A} Y' + e^- + \tilde{\nu}_e \qquad (\beta^-\text{-decay})$$

where X,X′ are 'parent' nuclei and Y,Y′ are 'daughter' nuclei with the charge and mass numbers indicated. There is also an associated process which competes with β^+-decay in which, instead of emitting a positron, the parent nucleus captures an electron from a (bound) K orbit and emits a neutrino, thus:

$$^A_Z X + e^-_k \rightarrow _{Z-1}^{A} Y + \nu_e$$

Although only electrons were observed in early studies of β^--decay it is postulated above that the electron is accompanied by a neutral particle known as an antineutrino and, similarly, the positron is accompanied by a neutrino – also neutral. This fundamental feature of β-decay was first proposed by Pauli in 1930 for two reasons. First, since the parent and daughter nuclei have well-defined energy levels between which the transition takes place, the energy released in a decay should have a definite value, say, E_0. Thus, in β^--decay,

---

*An antiparticle is one which has the same mass and spin as the corresponding particle but opposite charge. In the case of fermions it also has opposite signs for all other properties for which it is physically meaningful to reverse the sing, for example magnetic moment (in relation to the direction of spin) and parity. Examples of particle and antiparticle are electron (e^-) and positron (e^+) or proton and antiproton, the latter having negative charge. All particles have corresponding antiparticles although for a few electrically neutral particles (e.g. the photon) particle and antiparticle are identical.)

because of energy conservation, if only an electron is emitted it should have total energy (including mass energy) E_0. Experimentally, however, it is found that the emitted electrons have a continuous energy spectrum (Fig. 6.8) up to a maximum given by E_0. This can be explained, however, if E_0 is shared by two particles which can individually have a continuum of energies subject to the sum being equal to E_0. But, further, insofar as the maximum energy of the electrons (including their mass energy) is observed to be equal to E_0, this suggests that the mass of the antineutrino is very small, if not zero, otherwise the maximum electron energy would be E_0 less the rest mass energy of the antineutrino. The issue of antineutrino (and, correspondingly, neutrino) mass is of considerable topical importance and will be referred to later. (section 6.5.4)

The second reason for requiring the emission of an additional particle is to conserve angular momentum. Since A does not change in β-decay, it follows that the parent and daughter nuclei both have either integer or half-integer spin depending on whether A is even or odd (see section 2.5). In either case, if angular momentum is conserved, the decay product(s) must carry away an integer value of angular momentum. This is clearly not possible if only an electron (or positron), with $J = \frac{1}{2}$, is emitted. However, if the accompanying antineutrino (or neutrino) also has $J = \frac{1}{2}$ then there is no problem. As will be discussed later in this section the role of the neutrino (antineutrino) in β-decay with $m_v \simeq 0$ and $J = \frac{1}{2}$ is now well established.

It will be noted in the three β-decay processes that, although Z changes by ± 1, A remains constant; effectively a proton transforms into a neutron in β^+-decay and electron capture and a neutron into a proton in β^--decay. The basic transformations taking place, therefore, are

$$p \rightarrow n + e^+ + v_e \qquad \text{(in } \beta^+\text{-decay)}$$
$$n \rightarrow p + e^- + \bar{v}_e \qquad \text{(in } \beta^-\text{-decay)}$$
$$p + e_k^- \rightarrow n + v_e \qquad \text{(in electron capture)}$$

and these can be represented symbolically by the three diagrams given in Fig. 6.6. These are equivalent to the photon diagrams (Fig. 6.2) and G_β measures the strength of the β-decay interaction, just as e measures the strength of the electromagnetic interaction. It is intimately related to the overall weak interaction coupling constant and in due course its value will be determined from experimental data. Although the first two processes can take place when the nucleons are

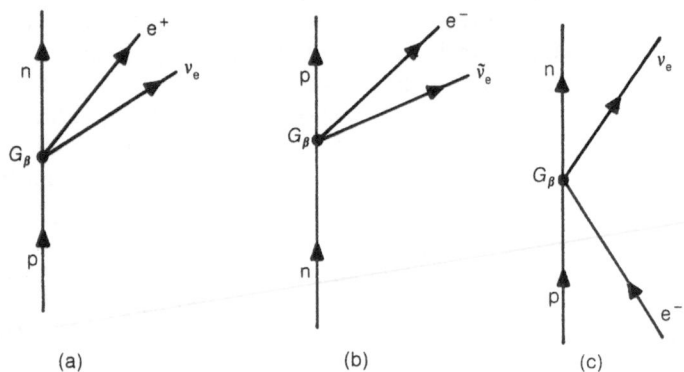

Fig. 6.6 Diagrammatic representation of (a) β^+-decay, (b) β^--decay and (c) electron capture.

contained in a nucleus, only the β^--decay of the neutron can take place in the free state. This is because the Q value (see below) for β^+-decay of the proton is negative.

For the three decay processes considered the Q values (discussed in section 5.1) are obtained very easily by considering the difference in mass energies between the initial and final states. Assuming that the ν_e and $\tilde{\nu}_e$ masses are zero we have for β^+-decay

$$Q_{\beta^+} = [M_{\mathrm{Nu}}(A, Z) - M_{\mathrm{Nu}}(A, Z - 1) - m_e]c^2$$

where M_{Nu} signifies the nuclear mass. Neglecting electron binding energy we can write for the atomic mass

$$M(A, Z) = M_{\mathrm{Nu}}(A, Z) + Zm_e \qquad (6.37)$$

Substituting in the above expression for Q_{β^+} then gives

$$Q_{\beta^+} = [M(A, Z) - M(A, Z - 1) - 2m_e]c^2 \qquad (6.38)$$

For β^--decay an equivalent calculation gives

$$Q_{\beta^-} = [M(A, Z) - M(A, Z + 1)]c^2 \qquad (6.39)$$

and for electron capture the result is

$$Q_{\mathrm{EC}} = [M(A, Z) - M(A, Z - 1)]c^2 \qquad (6.40)$$

For these different decay processes to take place the Q values must be positive and it will be noted that if β^+-decay can take place, then so can electron capture and the two processes compete. However, there are many decays for which Q_{EC} is positive but Q_{β^+} is negative (Fig. 6.7

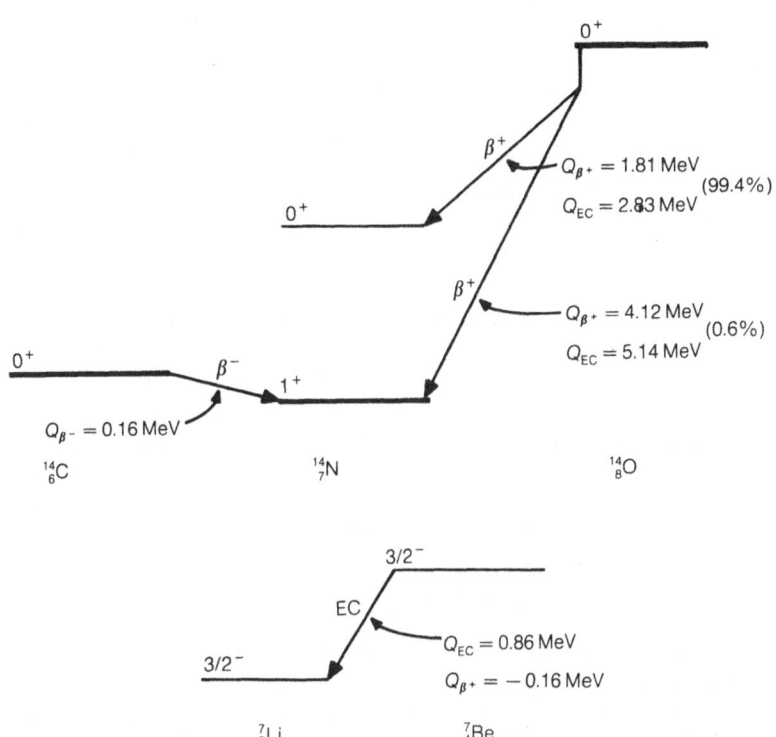

Fig. 6.7 Examples of β^--, β^+-decay and electron capture (EC). The vertical line included in β^+-decay symbolizes the additional energy $(2m_ec^2 = 1.022\,\text{MeV})$ needed for β^+-decay to take place compared with electron capture. 99.4% of the ^{14}O decays go to the first excited state of ^{14}N and only 0.6% to the ground state. The ^{7}Be decay is an example of a decay in which electron capture can take place, but not β^+-decay since Q_{β^+} is negative.

shows an example) so that the decay can only take place by electron capture. In the foregoing discussion it has been assumed that the decays take place to the ground state of the daughter nucleus. Clearly, for decays to excited states, the Q values should be reduced by the excitation energy of the state concerned. Examples of three typical beta decay processes are shown in Fig. 6.7.

6.5.1 The Beta-Decay Spectrum and Transition Probability

As with γ-decay transitions the probability per unit time of a β-decay transition is proportional to (coupling constant)^2 × (matrix element)^2, i.e. to $G_\beta^2 |M_\beta|^2$ where M_β is the nuclear β-decay matrix element. The nature of the latter will be discussed later; suffice it to say at this stage that it involves the wave functions of the initial and final states and contains all the specifically nuclear information about the decay. But in addition to these two factors another has to be included. This arises because two particles (e.g. e^+ and v_e) are emitted in the decay compared with one (the photon) in γ-decay.

The energy E_0 released in the β-decay is shared between these two particles and the probability of its being shared in a particular way depends on the number of states available to each particle together with the particular amount of energy it happens to have. To be more precise, consider β^--decay and let the magnitudes of the energy and momentum of the emitted electron be E and p respectively, and those of the antineutrino E_v and p_v. The following relationships then hold:

$$E = (p^2c^2 + m_e^2 c^4)^{\frac{1}{2}}$$
$$E_v = p_v c$$
$$E_0 = E + E_v \tag{6.41}$$

where, for the moment, it is assumed that the antineutrino has zero rest mass. The number of states available to an electron with momentum between p and $p + dp$ is proportional to $p^2 \, dp$, the usual element of volume in phase space, and similarly for the antineutrino. Thus the number of states available to both particles is proportional to the product of these two terms. However, the electron and

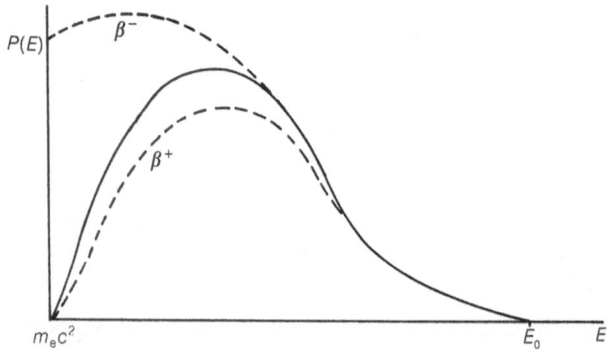

Fig. 6.8 β-decay spectrum shape. The full curve does not include Coulomb effects whilst the broken curves take them into account for β^+- and β^--decay.

antineutrino are not independent particles and this product must be multiplied by a further factor which ensures that it is only evaluated for values of p and p_v which are such that the sum of the corresponding energies equals E_0 (see the last of eqs (6.41)). Thus the number of states, dN, available to electron and antineutrino can be written

$$dN \propto p^2 \, dp \, p_v^2 \, dp_v \, \delta(E + E_v - E_0) \qquad (6.42)$$

Differentiating the first two of eqs (6.41) gives $E \, dE = c^2 p \, dp$ and $dE_v = c \, dp_v$ and substituting for dp and dp_v from these relationships gives

$$dN \propto pE \, dE \, E_v^2 \, dE_v \, \delta(E + E_v - E_0) \qquad (6.43)$$

In the above expressions δ is a Dirac δ-function which vanishes unless its argument is zero, i.e. unless the energy condition is satisfied*.

Since our concern is with the electron spectrum, the expression for dN should be integrated over all antineutrino energies. Using the properties of the δ-function outlined above, this gives at once for the number of electron states, with energy between E and $E + dE$,

$$dN_e \propto pE(E_0 - E)^2 \, dE \qquad (6.44)$$

Incorporating the coupling constant and nuclear matrix element terms referred to earlier gives finally for the probability per unit time of the emission of an electron in the energy range from E to $E + dE$:

$$P(E) \, dE \propto G_\beta^2 |M_\beta|^2 (E^2 - m_e^2 c^4)^{1/2} E(E_0 - E)^2 \, dE \qquad (6.45)$$

where the first of eqs (6.41) has been used to write p in terms of E.

A plot of $P(E)$ against E, assuming M_β is energy independent, is shown in Fig. 6.8 (full curve) and this is known as the statistical shape. It does, however, take no account of the electric field experienced by the electron as it escapes from the nucleus (charge Ze). Because of the attractive nature of this field, electrons are 'held back' and so the

---

*($\delta(x)$ is a function which has the value 0 for $x \neq 0$ and $\to \infty$ for $x = 0$ in such a way that

$$\int_{-\infty}^{+\infty} \delta(x) \, dx = 1$$

and

$$\int_{-\infty}^{+\infty} f(x)\delta(x - x_0) \, dx = f(x_0)$$

are satisfied.)

proportion of low energy electrons in the spectrum is increased as well as the total transition probability (determined by the integral of $P(E)$ – eq. (6.47)). Similarly, for β^+-decay, positrons are accelerated away and so the proportion of low energy positrons is decreased. These effects are shown in Fig. 6.8 (broken curves). Their calculation, using relativistic wave functions for electrons or positrons moving in the Coulomb field of the nucleus, is essentially straightforward, if a little tedious, and results in the inclusion of a further factor $F(Z, E)$, known as the Fermi function, in eq. (6.45). Finally, the constant of proportionality should be included, and this can be obtained by carrying out a full time-dependent perturbation theory calculation. The end result, writing $E = \varepsilon m_e c^2$, etc. is

$$P(\varepsilon)\,d\varepsilon = \frac{G_\beta^2}{2\pi^3}\frac{m_e^5 c^4}{\hbar^7}|M_\beta|^2(\varepsilon^2 - 1)^{1/2}\varepsilon(\varepsilon_0 - \varepsilon)^2 F(Z,\varepsilon)\,d\varepsilon \qquad (6.46)$$

For reasons which will be discussed in the next paragraph this is known as the 'allowed' spectrum shape.

The number $N(\varepsilon)$ of electrons emitted with energy ε is, of course, proportional to $P(\varepsilon)$ and, from eq. (6.46), it follows that a plot of $[N(\varepsilon)/(\varepsilon^2 - 1)^{1/2}\varepsilon F(Z,\varepsilon)]^{1/2}$ against ε should be a straight line intersecting the ε-axis at $\varepsilon = \varepsilon_0$. This is known as a Kurie plot and many β-decays are found to fit this diagram. It is clearly a way of establishing the energy E_0 released in a β-decay.

Throughout this discussion it has been assumed that $m_v = 0$. It is straightforward to calculate the decay spectrum without this approximation (problem 6.4) and the effect is to make the following replacement in the expression for $P(\varepsilon)\,d\varepsilon$ in eq. (6.46):

$$(\varepsilon_0 - \varepsilon)^2 \rightarrow [(\varepsilon_0 - \varepsilon)^2 - m_v^2/m_e^2]^{1/2}(\varepsilon_0 - \varepsilon)$$

Clearly the neutrino mass term, m_v, will only affect the spectrum significantly when $\varepsilon \simeq \varepsilon_0$ and this is a region where the intensity is very low (Fig. 6.8). Experiments are therefore very difficult and the present situation (1990) from studies of the end-point of the spectrum in the decay process $^3\text{H} \rightarrow {}^3\text{He} + \text{e}^- + \tilde{v}_e$ is that $m_v c^2 \leqslant 10\text{–}20\,\text{eV}$; very small compared with m_e ($\simeq 500\,000\,\text{eV}$).

The total transition probability, T_β, for a β-decay is obtained by integrating $P(\varepsilon)\,d\varepsilon$ over all possible energies of the emitted electron or positron, i.e. over the range from 1 to ε_0, thus:

$$T_\beta = \int_1^{\varepsilon_0} P(\varepsilon)\,d\varepsilon = \frac{G_\beta^2}{2\pi^3}|M_\beta|^2\frac{m_e^5 c^4}{\hbar^7}f(Z,\varepsilon_0) \qquad (6.47)$$

where

$$f(Z, \varepsilon_0) = \int_1^{\varepsilon_0} F(Z, \varepsilon)(\varepsilon_0 - \varepsilon)^2 (\varepsilon^2 - 1)^{1/2} \varepsilon \, d\varepsilon$$

The integration has to be performed numerically and tables of $f(Z, \varepsilon_0)$ for different values of Z and ε_0 are available.

The β-decay half-life $t_{1/2}$ (eq. (6.8)) is then given by

$$t_{1/2} = \frac{\ln 2}{T_\beta} = \frac{K}{f(Z, \varepsilon_0) G_\beta^2 |M_\beta|^2}$$

or

$$f(Z, \varepsilon_0) t_{1/2} = \frac{K}{G_\beta^2 |M_\beta|^2} \qquad (6.48)$$

where

$$K = \frac{2\pi^3 (\ln 2) \hbar^7}{m_e^5 c^4} = 1.23 \times 10^{-120} \, \text{J}^2 \, \text{m}^6 \, \text{s}$$

$f(Z, \varepsilon_0) t_{1/2}$ or, as normally written, ft, is known as the comparative half-life or ft value. The half-life can be measured experimentally and when multiplied by f (obtained from tables) the resulting ft value can be used to determine information about the magnitudes of G_β and M_β. This will be our concern in the next section.

For electron capture, only one particle (ν_e) is emitted and so it has a unique energy Q_{EC} (eq. (6.40)). Taking K-electron capture as an example, the probability of capture per unit time is proportional to three factors: (i) $G_\beta^2 |M_\beta|^2$ as with $\beta^\pm$-decay, (ii) the probability, $|\psi_K(0)|^2$, of finding a K-electron in the nucleus, where $\psi_K(r)$ is its wave function and (iii) the density of states available to the neutrino which, following earlier discussion, is proportional to $p_\nu^2 \propto Q_{EC}^2$. Thus

$$T_K \propto G_\beta^2 |M_\beta|^2 |\psi_K(0)|^2 Q_{EC}^2$$

For a K-electron the hydrogen-like wave function, ψ_K, is well known and, for $r = 0$, is proportional to $(Z/a_0)^{3/2}$ where a_0 is the Bohr radius. Substituting for this and taking account of the factors arising in a full calculation the following result is obtained:

$$T_K = \frac{G_\beta^2}{2\pi^3} |M_\beta|^2 \frac{2\pi m_e}{c\hbar^4} \left(\frac{Z}{a_0}\right)^3 Q_{EC}^2 \qquad (6.49)$$

In cases where K-capture competes with β^+-decay it can be seen by comparing eqs (6.47) and (6.49) that the ratio of transition probabilities is independent of the coupling constant and the matrix element. Roughly speaking it is found that K-capture is most important for large Z values and β^+-decay for large values of ε_0. The process of K-capture is signalled by the emission of K X-rays or Auger electrons resulting from the filling of the vacancy in the K shell by another atomic electron. L-, M-,... captures can also take place but are much less probable processes.

6.5.2 The Nuclear Physics of Beta-Decay

The above discussion focused on the β-decay energy spectrum whose shape is largely determined by statistical considerations and the nuclear Coulomb field. The intrinsic nuclear physics of the decay process has so far been subsumed into the matrix element M_β and this must now be studied.

Consider first the orbital angular momentum carried away by the emitted particles. Typical decay energies are of the order of 1 MeV corresponding to a momentum for an electron or neutrino of this energy of the order $p \simeq 5 \times 10^{-22}\,\mathrm{J\,m^{-1}\,s}$. If the emitted particles carry away angular momentum $\simeq L\hbar (L = 0, 1, 2, \ldots)$ they will need to be emitted a distance $\simeq L\hbar/p \simeq 2L \times 10^{-13}\,\mathrm{m}$ from the centre of the nucleus. For non-zero L this is large compared with typical nuclear radii ($\simeq 10^{-14}\,\mathrm{m}$) and is therefore in a region of low nuclear density. This implies that unless $L = 0$ the β-decay has a low probability of occurring. For this reason, decays for which $L = 0$ are said to be **allowed** and the rest **forbidden** ($L = 1$, first forbidden; $L = 2$, second forbidden, etc.). Whilst allowed transitions have ft values in the range 10^3–10^5 s, forbidden transitions have ft values $\geqslant 10^6$ s. We shall now confine our attention to allowed transitions.

The emitted particles (e.g. e^- and $\tilde{\nu}_e$) each have spin $\frac{1}{2}$ and therefore can be in a singlet ($S = 0$) or triplet ($S = 1$) state. In the former case the β-decay process is referred to as **Fermi** or **polar vector** (V) transition and in the latter as a **Gamow–Teller** or **axial vector** (A) transition. The former terminology is historical and the latter refers to the detailed mathematical form of the fundamental weak interaction responsible for β-decay which will be discussed in Chapter 9.

For allowed transitions since $L = 0$ (even parity) the initial and final nuclear states must have the same parity; there is no parity change. However, for a Fermi transition, since $S = 0$, no angular momentum is carried away whilst for a Gamow–Teller transition, with $S = 1$, one unit of angular momentum is removed. Thus, using angular momentum conservation, the nuclear spin–parity selection rules for these two types of allowed decay can be written:

Fermi transition (V) $\qquad\qquad \Delta J = 0,$ $\qquad\qquad\qquad \Delta P = $ 'no'
Gamow–Teller transition (A) $\quad \Delta J = 0, \pm 1 \,(0 \nrightarrow 0),$ $\qquad \Delta P = $ 'no'

Although we shall not be using them, selection rules for forbidden transitions can be derived in a similar fashion (problem 6.5).

The above selection rules follow mathematically from the matrix element M_β which vanishes for an allowed transition unless the selection rules are satisfied. To understand this the following replacement should be made in the expression for ft (eq. (6.48)):

$$G_\beta^2 |M_\beta|^2 \rightarrow G_V^2 |M_V|^2 + G_A^2 |M_A|^2 \qquad (6.50)$$

where the modern V and A notation is now used to indicate Fermi and Gamow–Teller transitions respectively and two different coupling constants have been introduced corresponding to the two forms of transition. Equation (6.48) can now be written

$$ft = \frac{K}{G_V^2 |M_V|^2 + G_A^2 |M_A|^2} \qquad (6.51)$$

Explicitly M_V and M_A are given by

$$M_V = \int u_f^* (\sum t_j^\pm) u_i \, d\tau, \qquad M_A = \int u_f^* (\sum \sigma_j t_j^\pm) u_i \, d\tau$$

where $t^\pm$ is an operator which transforms a neutron into a proton ($+$) or a proton into a neutron ($-$) and σ is the usual Pauli spin operator, the sum being taken over all nucleons in the nucleus.

The task of evaluating these matrix elements depends of course on detailed knowledge of the nuclear wave functions. There are, however, two simple and important cases to be considered. The first is exemplified by the $0^+ \rightarrow 0^+$ β^+ transition between ^{14}O and the first excited state of ^{14}N shown in Fig. 6.7. From the above selection rules this can only be a Fermi (V) transition ($M_A = 0$) and further it is taking place between two members of an isospin multiplet (section 3.3.1). The two states are therefore essentially identical except that a proton in ^{14}O has been replaced by a neutron in ^{14}N. Because of this M_V is independent of nuclear structure details and simply has the value $\sqrt{2}$. This is a relatively large value and a decay such as this is known as a **superallowed transition**. In general, because of fundamental differences in wave functions of initial and final states, matrix elements are much smaller than 1. Substituting $M_V = \sqrt{2}$ and $M_A = 0$

in eq. (6.51) then gives

$$ft = \frac{K}{2G_V^2}$$

The measured ft value is $ft = 3037.1 \pm 1.6$ s and, using the value for K in eq. (6.48), the value $G_V \simeq 1.42 \times 10^{-62}$ J m^3 is obtained. Taking into account many small electromagnetic and nuclear corrections and the experimental data from other superallowed transitions the following more precise value results:

$$G_V = (1.397 \pm 0.001) \times 10^{-62} \text{ J m}^3 \qquad (6.52)$$

The other simple decay to consider is that of the neutron itself, since here there are no nuclear complexities. Since this is a $\frac{1}{2}^+ \rightarrow \frac{1}{2}^+$ transition there can be both V and A contributions and, after allowing for the different possible initial and final spin orientations of the neutron and proton, the two matrix elements are found to have the values $M_V = 1$, $|M_A| = \sqrt{3}$. Measurement of the neutron lifetime is difficult, involving the use of neutron beams from a reactor or 'bottled' (confined by electric and magnetic fields) neutrons. A recent value for the neutron half-life is $t_{1/2} = 623.6 \pm 6.2$ s with a corresponding ft value, including some small electromagnetic corrections, $ft = 1063 \pm 21$ s. Using this in eq. (6.50) together with the above values of the matrix elements and the value for G_V given in eq. (6.51) enables G_A to be determined. The result is

$$|G_A| = (1.781 \pm 0.011) \times 10^{-62} \text{ J m}^3 \qquad (6.53)$$

so that if $\lambda = G_A/G_V$, $|\lambda| \simeq 1.27$. The sign of λ is not determined by these measurements. However, the angular distribution of electrons in the decay of polarized neutrons depends on interference between the V and A terms so that their relative signs can be determined. Experiment establishes that λ is, in fact, negative. For this reason the β-decay interaction is generally referred to as a V–A interaction.

Later, in Chapter 9, we shall see that the values of G_V and G_A play an important part in discussing the general properties of the weak interaction from which they derive. Here we just note that if G_V is written in dimensionless form, i.e. $G_V(m_p c^2)^{-1} (\hbar/m_p c)^{-3}$, a value of the order 10^{-5} is obtained. This is to be compared with the dimensionless form of the strong and electromagnetic coupling constants, namely $\alpha_s = g_s^2/\hbar c \simeq 1$ (section 3.4) and $\alpha = e^2/4\pi\varepsilon_0 \hbar c \simeq 1/137$ respectively. The comparison indicates the relative weakness of the β-decay interaction.

Many other allowed and forbidden transitions have been studied theoretically and experimentally using eq. (6.50) for the ft value. Knowing the two coupling constants, the thrust of these studies has been to interpret experimental data on lifetimes by calculating the matrix elements using wave functions generated by different nuclear models. Much success and understanding has been achieved.

6.5.3 Parity Non-Conservation

So far it has been assumed that parity is conserved in all nuclear processes. This derives from the assumption that the nuclear and interaction Hamiltonians, denoted by H, are invariant under the operation of reflecting all spatial coordinates through the origin – sometimes loosely referred to as 'mirror reflection' and denoted symbolically by $r \rightarrow -r$ (a vector such as r which changes sign under reflection of axes is referred to as a **polar** vector). Thus

$$H(r) = H(-r)$$

H is said to be a **scalar** quantity as distinct from a **pseudoscalar** which changes sign under the above operation. This invariance carries with it the implication that all energy eigenstates have a definite parity (even or odd) and that the total (multiplicative) initial and final parities in any transition should be the same – parity is conserved. In particular all transition probabilities should be scalar quantities. In terms of 'mirror' language this means that the mirror image of any physical process for which parity is conserved should also be an allowed physical process. This then excludes processes which exhibit an intrinsic (left or right) 'handedness' because the mirror image would have the opposite handedness and would, therefore, not be allowed.

In 1956 it was suggested by Lee and Yang, for reasons that will be mentioned in section 7.3.2, that the weak interaction and β-decay in particular did not conserve parity. Thus the β-decay interaction Hamiltonian, H_β, was conjectured to have both scalar (S) and pseudoscalar (P) components, i.e.

$$H_\beta = H_{\beta S} + H_{\beta P}$$

In turn the overall transition matrix element, $\mathfrak{M}_\beta$, which, in addition to the nuclear part M_β introduced earlier in this section, also includes a part involving electron and neutrino wave functions, has scalar and pseudoscalar components, i.e.

$$\mathfrak{M}_\beta = \mathfrak{M}_{\beta S} + \mathfrak{M}_{\beta P}$$

The transition probability, being proportional to $|\mathfrak{M}_\beta|^2$, will therefore involve a cross-term of a scalar and a pseudoscalar which is itself a pseudoscalar. It is this term which could lead to a handedness in the β-decay process.

In 1957, Wu *et al.* carried out an experiment which studied the angular distribution of β-decay electrons (momentum p) in relation to the spin (J) of the β-decaying nucleus. If parity is not conserved in the process it was predicted that the angular distribution should have the form

$$I(\theta) \propto 1 + a\hat{\boldsymbol{J}}\cdot\hat{\boldsymbol{p}} = 1 + \alpha\cos\theta \qquad (6.54)$$

where α depends on the details of the β-decay interaction, $\hat{\boldsymbol{J}}$ and $\hat{\boldsymbol{p}}$ are unit vectors in the directions of J and p respectively and θ is the angle between them. J is an angular momentum vector ($\sim r \times p$) and remains the same under reflection of axes ($r \to -r$, $p \to -p$); it is referred to as an **axial** vector. Since p is a polar vector, the product $\hat{\boldsymbol{J}}\cdot\hat{\boldsymbol{p}}$ is the expected pseudoscalar.

The actual experiment used the decay process

$$^{60}\text{Co}(J=5) \to {}^{60}\text{Ni}(J=4) + \text{e}^- + \tilde{\nu}_\text{e}$$

The spins of ^{60}Co were lined up (**polarized**) and a measurement made of the 'up–down' asymmetry of the emitted electrons in relation to the polarization direction. The polarization was effected by subjecting the nuclei to a powerful magnetic field, which interacted with their magnetic moments, and maintaining them at a temperature of $\sim 0.01\,\text{K}$ so as to reduce perturbing thermal fluctuations to a minimum. It was found that electrons were emitted preferentially in the opposite direction to the nuclear spin (Fig. 6.9). The results of this experiment and of subsequent more accurate experiments demonstrate unequivocally that parity is not conserved in β-decay. They are consistent with the value of the coefficient $\alpha = -v/c$ where v is the electron speed.

Another result follows from this experiment, namely that the electrons are mostly spinning in a left-handed fashion about their direction of motion. This follows from angular momentum conservation considerations. The transition is an allowed Gamow–Teller (A) transition and so the emitted leptons have their spins parallel ($S=1$) and to conserve angular momentum S must be parallel to the spin $J=5$ of the decaying nucleus (Fig. 6.9). However, the electrons are emitted predominantly in the opposite direction to this nuclear spin and must therefore be mostly spinning in a left-handed fashion. To be more precise we define the **helicity** (λ) of the electrons as the average of

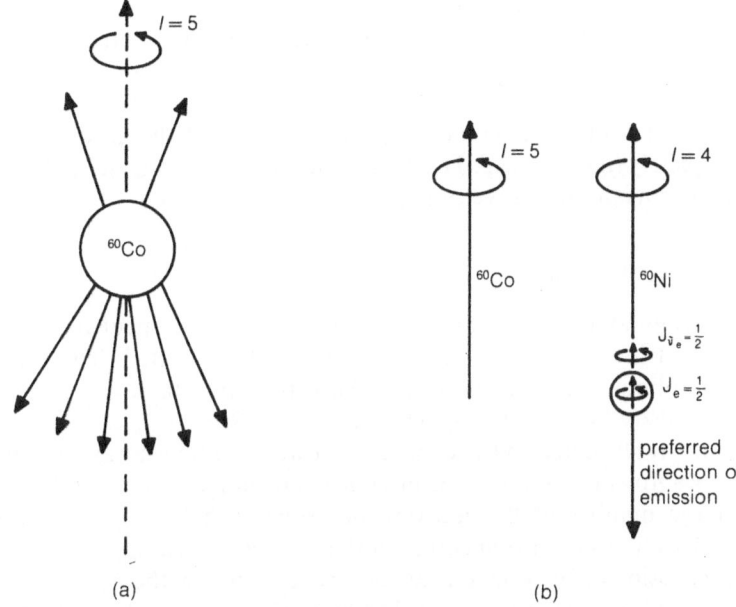

Fig. 6.9 Parity non-conservation in the β^--decay of $^{60}\text{Co} \rightarrow {}^{60}\text{Ni}$: (a) preferential emission of electrons opposite to the nuclear spin direction; (b) conservation of angular momentum requires the spin of the electron to be left-handed in relation to its direction of motion.

the projection of their spin direction $(\hat{\boldsymbol{J}}_\text{e})$ along the direction of their momentum $(\hat{\boldsymbol{p}})$, i.e.

$$\lambda = (\hat{\boldsymbol{J}}_\text{e} \cdot \hat{\boldsymbol{p}})_\text{av} \tag{6.55}$$

Detailed experiments, for example in which β^--decay electrons are scattered by other electrons in a magnetized foil, have established that they have $\lambda = -v/c$. Conversely, positrons emitted in β^+-decay have $\lambda = +v/c$; they are right-handed. Measurements of neutrino and antineutrino helicities have also been made, in the first instance by Goldhaber *et al.* in 1958. They measured the circular polarization of photons emitted following electron capture in the sequential processes

$$^{152}\text{Eu} + \text{e}^- \rightarrow {}^{152}\text{Sm}^* + \nu_\text{e}$$
$$^{152}\text{Sm}^* \rightarrow {}^{152}\text{Sm} + \gamma$$

This circular polarization is intimately related by angular momentum conservation to the neutrino helicity and their results coupled with those from subsequent experiments have established that

$$\lambda(v_e) = -1; \qquad \lambda(\tilde{v}_e) = 1 \qquad (6.56)$$

Thus the neutrino is completely left-handed and the antineutrino completely right-handed. This handedness of the leptons emitted in β-decay is again indicative of parity non-conservation.

6.5.4 The Neutrino

Throughout the previous discussion the role of neutrinos and antineutrinos in β-decay has been taken for granted. The basic assumptions and deductions have been that $m_v \simeq 0$, $J_v = \frac{1}{2}$, $\lambda(v_e) = -1$ and $\lambda(\tilde{v}_e) = +1$. It was, in fact, not until 1953 that these particles were actually detected by Reines and Cowan in a β-decay experiment using antineutrinos produced in a nuclear reactor. In a reactor there is a large number of β$^-$-decaying neutrons each having a half-life $\simeq 10.5$ min (as discussed earlier in this section). Each decay involves the emission of an antineutrino and so a large flux of these particles is produced. These can be detected by using the inverse β-decay process (a variant on β$^+$-decay)

$$\tilde{v}_e + p \rightarrow n + e^+$$

The experiment used a large container of hydrogenous material adjacent to the reactor to which some cadmium had been added and in which both the neutron and the positron could be detected. The former was detected by looking for de-excitation γ-rays following its capture by cadmium, which has a very large neutron capture cross-section, and the latter by looking for annihilation radiation ($e^+ + e^- \rightarrow 2\gamma$) using scintillating material as a γ-detector. The cross-section for the reaction is estimated to be very small ($\simeq 10^{-19}$ b) and only a few counts per hour were recorded. But the experiment clearly showed the physical presence of antineutrinos.

Another capture reaction that has been used is

$$v_e + {}^{37}\text{Cl} \rightarrow {}^{37}\text{Ar} + e^-$$

as a means of detecting neutrinos emitted in fusion processes (section 5.11) taking place in the sun. Suffice it to say that, although the reaction has been observed, the implied flux of solar neutrinos is significantly less than predicted by current theories of stellar structure. The above reaction has also been studied using an incoming flux of antineutrinos from a reactor and no capture was observed. This implies that the neutrino and antineutrino are fundamentally different particles which is consistent with the fact that they have already been deduced to have opposite helicities. Note that if the helicities are exactly $\lambda = \pm 1$ then $m_v = 0$. This follows because if $m_v \neq 0$ its speed must be less than c. This means that a $\lambda = -1$ neutrino could be 'overtaken' and observed in a frame of reference in which it would appear to be moving away from the observer with $\lambda = +1$.

But, as has already been mentioned, there is still uncertainty about the mass of the neutrino. It is certainly very small ($\leqslant 10 - 20 \, \text{eV}$) and may be zero. Knowledge of its precise value is of vital importance to astronomers and experiments to measure it more exactly are underway.

6.5.5 Double Beta Decay

One process that may help with the elucidation of neutrino properties is what is known as **double β-decay**. Sometimes there are situations when a nucleus is energetically stable against ordinary β-decay but can decay with the emission of two β-particles. A case in point is $^{82}_{34}\text{Se}$ which, because it is tightly bound, being an even–even nucleus, cannot β-decay to $^{82}_{35}\text{Br}$ but, with the emission of two electrons, can double β-decay to the equally tightly bound even–even nucleus $^{82}_{36}\text{Kr}$.

There are, in principle, two ways in which the decay can take place, namely

$$^{82}\text{Se} \rightarrow {}^{82}\text{Kr} + 2e^- + 2\tilde{v}_e$$

or

$$^{82}\text{Se} \rightarrow {}^{82}\text{Kr} + 2e^-$$

The first process is simply 'double action' of the ordinary β-decay process but the second, in which no neutrinos are emitted, is quite different and can only occur via the ordinary β-decay interaction if it proceeds in two stages:

$$^{82}\text{Se} \rightarrow {}^{82}\text{Br} + e^- + \tilde{v}_e$$
$$\tilde{v}_e + {}^{82}\text{Br} \rightarrow {}^{82}\text{Kr} + e^-$$

where the first is a virtual process, not being allowed energetically, and in the second the $\tilde{v}_e$ emitted in the first stage is absorbed by the virtual ^{82}Br, the total outcome being the emission of two electrons only. The second stage is, of course, contrary to the findings about $\tilde{v}_e$

absorption mentioned above and would imply that the ν_e and $\tilde{\nu}_e$ were not completely different particles with $\lambda = \pm 1$ and, further, would have non-zero mass. Under these circumstances they are called Majorana neutrinos as distinct from Dirac neutrinos if they are completely different.

Clearly a search for neutrinoless double β-decay is important and so far it has been unsuccessful. A direct laboratory measurement of ^{82}Se decay has given a half-life of about 1.1×10^{20} years and this is consistent with calculations for the half-life assuming the emission of antineutrinos. Further, if no antineutrinos are emitted, the total energy of the two electrons should be equal to the energy released in the decay, so leading to a 'spike' in the energy distribution. Such a spike has not been observed. Other investigations, for example, of the decay $^{130}_{52}$Te $\rightarrow$ $^{130}_{54}$Xe in which measurements are made in natural ores of the abundance of parent and daughter nuclei lead to half-life estimates of the order 10^{21} years and are again consistent with antineutrino emission. Very recently, a search for a spike in the double β-decay ^{76}Ge $\rightarrow$ ^{76}Se has established that the half-life for neutrinoless double β-decay, if it takes place, must be greater than 2.3×10^{24} years which, in turn, implies that $m_\nu c^2 < 3\,\mathrm{eV}$. The search continues.

7

Elementary particles and their interactions

7.1 INTRODUCTION

In the foregoing discussion of nuclear phenomena a number of elementary particles have been encountered and some features of their interactions have been described. Each of these particles can be characterized by its spin J, its electric charge (usually denoted by superfixes $+$, 0, $-$) and the interactions – strong, electromagnetic, weak – it experiences. These particles include:

1. the neutron and proton (n, p; $J = \frac{1}{2}$) which experience all three basic interactions;
2. the pions (π^+, π^0, π^-; $J = 0$) which experience all three basic interactions;
3. the electron and positron (e^+, e^-; $J = \frac{1}{2}$) which experience the electromagnetic and weak interactions;
4. the neutrino and antineutrino (ν_e, $\tilde{\nu}_e$; $J = \frac{1}{2}$) which experience only the weak interaction;
5. the photon (γ; $J = 1$) which experiences only the electromagnetic interaction.

We shall see in due course that these particles form the tip of an iceberg and that they are simply the lightest members of extended families of elementary particles. These different particles are generally produced in collisions between a high energy incident particle produced in some form of accelerator (section 5.3.1) and a target particle or nucleus. The higher the energy of the incident particle is, the heavier are the particles that can be produced in the process. Typically in current experiments protons are accelerated up to energies in the region of hundreds of GeV ($\sim 10^{11}$ eV) and they collide either with a stationary target or with antiprotons which have been accelerated to the same energy in the same accelerator but moving in the opposite direction – a 'colliding beam' experiment. In the latter case, since there is no net centre-of-mass motion (section 5.1) far more energy is available to create product particles. High energy electrons are also used in colliding beam experiments with positrons and, very recently, with protons. It is through studies of this kind that new particles are created either directly through the initial collision or indirectly as decay products of the different particles produced in the collision.

Collectively those particles which experience the strong interaction are referred to as **hadrons** and this large family is divided into **baryons** (fermions, with $J =$ half-integer, such as the neutron and proton) and **mesons** (bosons, with $J =$ integer, such as the pions). As has already been mentioned the $e^{\pm}$, v_e,... are referred to as **leptons**. Finally, particles such as the photon, which propagate a basic interaction (in this case the electromagnetic interaction), are known as **gauge bosons.**

In concluding this introductory resumé the reader is reminded that all particles have a corresponding antiparticle. This correspondence derives essentially from the relativistic relation between energy, rest mass and momentum, i.e.

$$E^2 = m_0^2 c^4 + p^2 c^2 \tag{7.1}$$

which allows mathematically both a positive and a negative value for E. This double-valuedness led Dirac in 1928 to propose the existence of the positron (e^+) as a companion antiparticle to the electron (e^-), a proposal which was confirmed experimentally by Anderson a few years later. Since then many particle–antiparticle pairs have been established which have the same mass but opposite charge and magnetic moment. In the case of fermions, particle and antiparticle can only be produced together (**pair production**) and they also annihilate each other, their total mass energy being carried away by some other form of radiation, e.g.

$$e^+ + e^- \rightarrow 2\gamma$$

$$p + \tilde{p} \rightarrow \text{several } \pi s$$

For some neutral bosons, particle and antiparticle are identical (e.g. $\tilde{\gamma} = \gamma$, $\tilde{\pi}^0 = \pi^0$); they are said to be **self-conjugate.**

7.2 TABLES OF ELEMENTARY PARTICLES

Over the years many hundreds of elementary particles have been identified with a very wide range of masses. They are virtually all unstable and so can generally be characterized by a mean life (τ) or, alternatively a width ($\Gamma \simeq 1/\tau$ – section 6.1). The latter is generally used when the lifetime is very short. Extensive tables are regularly compiled (e.g. Particle Data Group, 1988) listing particle properties such as spin (J), mass, charge, lifetime and decay modes, and in Table 7.1 an abbreviated version is given of such tables for a selection of particles which will form the basis of our subsequent discussion.

In the table, approximate value of the mass is given in MeV/c^2. Three properties given in the table – **isospin** (I), **strangeness** (S) and **intrinsic parity** (denoted by a superfix, P, to J) – have not yet been discussed and will be introduced in section 7.3. The different decay modes of the particles list the particles emitted in a particular mode (e.g. pe$^-$ $\bar{\nu}_e$ in the β-decay of the neutron) together with a percentage figure, if there is more than one mode, indicating the fraction of decays (the branching fraction) by this mode; errors are generally not included.

7.3 INTRINSIC PARTICLE PROPERTIES AND CONSERVATION LAWS

The different particle properties listed in Table 7.1, and additional properties which will be discussed in the following, are all associated with conservation laws. Some laws apply to all processes (e.g. conservation of energy, angular momentum and electric charge) whilst others only apply to certain interaction processes (e.g. parity conservation does not apply to weak processes). In some cases a conservation law relates to the invariance of the relevant Hamiltonian under a particular transformation. For example, conservation of angular momentum follows from the invariance of all Hamiltonians under spatial rotation and we have seen (section 6.5.3) that parity conservation follows from the invariance of a Hamiltonian under reflection of all spatial coordinates. Charge conservation, baryon and lepton number conservation follow from invariance under what is known as a **gauge transformation**. Some laws are additive in form, i.e. in a particular process 'sum of initial properties' = 'sum of final properties' (e.g. energy or charge conservation). Others are multiplicative, i.e. 'product of initial properties' = 'product of final properties' (e.g. parity conservation).

Table 7.1 Elementary particles

	Gauge bosons ($J = 1$)		
Particle	Mass (MeV/c^2)	Width (GeV)	Decay modes and branching fractions
γ	$< 3 \times 10^{-33}$	stable	—
W	$(80.49 \pm 0.67) \times 10^3$	< 6.5	$e\nu_e$, 10%; $\mu\nu_\mu$, 12%; $\tau\nu_\tau$, 10%
Z	$(91.49 \pm 1.39) \times 10^3$	< 5.6	$e^+ e^-$, 4.6%

	Leptons ($J = \frac{1}{2}$)		
Particle	Mass (MeV/c^2)	Mean life (s)	Decay modes and branching fractions
ν_e	$< 1.0 \times 10^{-5}$	stable	—
e^-	$0.510\,999\,06 \pm 0.000\,000\,16$	$> 2 \times 10^{22}$ years	—
ν_μ	< 0.25	stable	—
μ^-	$105.658\,39 \pm 0.000\,06$	$(2.197\,03 \pm 0.000\,04) \times 10^{-6}$	$e^- \tilde{\nu}_e \nu_\mu$, 100%
ν_τ	< 35	—	—
τ^-	1784.2 ± 3.2	$(3.3 \pm 0.4) \times 10^{-13}$	$\mu^- \tilde{\nu}_\mu \nu_\tau$, (17.8 ± 0.4)%; $e^- \tilde{\nu}_e \nu_\tau$, (17.5 ± 0.4)%; $\pi^- \nu_\tau$, (10.8 ± 0.6)%; $\rho^- \nu_\tau$ (22.3 ± 1.1)%

Baryons ($J^P = \frac{1}{2}^+$)

Particle	I	S	Mass (MeV/c^2)	Mean life (s)	Decay modes and branching fractions
p	$\frac{1}{2}$	0	938.28	stable	—
n		0	939.57	899.7 ± 8.9	$pe^- \nu_e$, 100%
Λ	0	-1	1115.60	2.631×10^{-10}	$p\pi^-$, 64.2%; $n\pi^0$, 35.8%; $pe^- \bar{\nu}_e$, 8.3×10^{-4}
Σ^+		-1	1189.37	0.800×10^{-10}	$p\pi^0$, 51.64%; $n\pi^+$, 48.36%; $\Lambda e^+ \nu_e$, 2×10^{-5}
Σ^0	1	-1	1192.46	7.4×10^{-20}	$\Lambda\gamma$, $\sim 100\%$
Σ^-		-1	1197.34	1.479×10^{-10}	$n\pi^-$, $\sim 100\%$; $ne^- \bar{\nu}_e$, 1.02×10^{-3}; $\Lambda e^- \bar{\nu}_e$, 5.73×10^{-5}
Ξ^0	$\frac{1}{2}$	-2	1314.9	2.90×10^{-10}	$\Lambda\pi^0$, $\sim 100\%$
Ξ^-		-2	1321.32	1.64×10^{-10}	$\Lambda\pi$, $\sim 100\%$; $\Lambda e^- \bar{\nu}_e$, 5.5×10^{-4}

Charmed baryon

Particle	I	S	Mass (MeV/c^2)	Mean life (s)	Decay modes and branching fractions
Λ_c^+	0	0	2284.9	1.8×10^{-13}	$pK^- \pi^+$, 2.2%; Λ any, 33%

Table 7.1 (*Continued*)

Baryons ($J^P = \frac{3}{2}^+$)

Particle	I	S	Mass (MeV/c^2)	Width (MeV)	Decay modes and branching fractions
$\Delta(1232)$	$\frac{3}{2}$	0	1232.0	115	$(n)\pi$, 99.4%; $(n)\gamma$, 0.6%
$\Sigma(1385)^+$	1	-1	1382.8	36	$\Lambda\pi$, 88%; $\Sigma\pi$, 12%
$\Sigma(1385)^0$		-1	1383.7	36	
$\Sigma(1385)^-$		-1	1387.2	39	
$\Xi(1530)^0$	$\frac{1}{2}$	-2	1531.8	9.1	$\Xi\pi$, 100%
$\Xi(1530)^-$		-2	1535.0	9.9	
Ω^-	0	-3	1672.4	0.822×10^{-10} s (lifetime)	ΛK^-, 67.8%; $\Xi^0\pi^-$, 23.6%; $\Xi^-\pi^0$, 8.6%; $\Xi^0 e^- \bar{\nu}_e$, 5.6×10^{-3}

Mesons ($J^P = 0^-$)

Particle	I	S	Mass (MeV/c²)	Mean life or width	Decay modes and branching fractions
π^+	1	0	139.57	2.60×10^{-8} s	$\mu^+ \nu_\mu$, $\sim 100\%$; $e^+ \nu_e$, 1.23×10^{-4}
π^0		0	134.97	0.84×10^{-16} s	$\gamma\gamma$, 98.8%
π^-		0	139.57	2.60×10^{-8} s	$\mu^- \bar{\nu}_\mu$, $\sim 100\%$; $e^- \bar{\nu}_e$, 1.23×10^{-4}
η^0	0	0	548.88	1.08 keV	$\pi^+ \pi^0 \pi^-$, 23.7%; $3\pi^0$, 31.90%; $\gamma\gamma$, 38.9%
K^+	$\frac{1}{2}$	+1	493.65	1.237×10^{-8} s	$\mu^+ \nu_\mu$, 63.5%; $\pi^+ \pi^0$, 21.2%; $\pi^+ \pi^- \pi^-$, 5.6%; $\pi^0 e^+ \nu_e$, 4.8%
K^0		+1	497.67	$K_S^0\ 0.8928 \times 10^{-10}$ s $K_L^0\ 5.183 \times 10^{-8}$ s	$\pi^+ \pi^-$, 68.6%; $\pi^0 \pi^0$, 31.4% $\pi^0 \pi^0 \pi^0$, 21.7%; $\pi^+ \pi^0 \pi^-$, 12.4%; $\pi^\pm \mu^\mp \nu_\mu$, 27.1%; $\pi^\pm e^\mp \nu_e$, 38.7%
$\eta'(958)$	0	0	957.57	0.24 MeV	$\eta\pi\pi$, 65.2%; $\rho^0 \gamma$, 30.0%
$\eta_c(2980)$	0	0	2979.6	10.3 MeV	$\eta'\pi\pi$, 4.1%; $K\bar{K}\pi$, 5.5%
D^+	$\frac{1}{2}$	0	1869.3	10.7×10^{-13} s	e^+ any, 19.2%; $K^- \pi^+ \pi^+$, 7.8%; $K^- \pi^+ \pi^+ \pi^0$, 3.7%; $\bar{K}^0 \pi^+ \pi^0$, 8.3%; $\bar{K}^0 \pi^+ \pi^+ \pi^-$, 7.0%
D^0		0	1864.5	4.3×10^{-13} s	e^+ any, 8%; K^- any, 43%; $K^- \pi^+ \pi^0$, 12.5%; $K^- \pi^+ \pi^+ \pi^-$, 7.9%; $K^0 \pi^+ \pi^-$, 5.6%
D_s^+	0	-1	1969.3	4.4×10^{-13} s	$\phi\pi^+$; $\phi\pi^+ \pi^+ \pi^-$

Bottom mesons

Particle	I	S	Mass (MeV/c²)	Mean life or width	Decay modes and branching fractions
B^+	$\frac{1}{2}$	0	5277.6	1.4×10^{-12} s	$\bar{D}^0 \pi^+$, 0.5%; $D^*(2010)\pi^+ \pi^+ \pi^0$, 4.3%; ψ any, 1%
B^0		0	5275.2		$\bar{D}^0 \pi^+ \pi^-$, $< 3.9\%$; $D^*(2010)\pi^+$, 0.3%

Table 7.1 (*Continued*)

Mesons ($J^P = 1^-$)

Particle	I	S	Mass (MeV/c^2)	Width (MeV)	Decay modes and branching fractions
$\rho(770)$	1	0	770.0	153.0	$\pi\pi$, $\sim 100\%$; e^+e^-, 0.0044%
$\omega(783)$	0	0	782.0	8.5	$\pi^+\pi^-\pi^-$, 89.3%; $\pi^0\gamma$, 8.0%; e^+e^-, 0.0071%
$\phi(1020)$	0	0	1019.4	4.41	K^+K^-, 49.5%; K^0_S and K^0_L, 34.4%; $\pi^+\pi^-\pi^0$, 1.9%; e^+e^-, 0.031%
$\left.\begin{array}{l}K^{*+}(892)\\K^{*0}(892)\end{array}\right\}$	$\frac{1}{2}$	$+1$ $+1$	892.1	51.3	$K\pi$, $\sim 100\%$
$J/\psi(3097)$	0	0	3096.9	0.068	e^+e^-, 6.9%; hadrons + radiative, 86.2%; $\mu^+\mu^-$, 6.9%
$\Upsilon(9460)$	0	0	9460.3	0.052	e^+e^-, 2.5%; $\tau^+\tau^-$, 3.0%; $\mu^+\mu^-$, 2.6%
$\left.\begin{array}{l}D^{*+}(2010)\\D^{*0}(2010)\end{array}\right\}$	$\frac{1}{2}$	0 0	2010.1 2007.1	<2.0 <5	$D^0\pi^+$, 49%; $D^+\pi^0$, 34% $D^0\pi^0$, 52%; $D^0\gamma$, 48%

In the following discussion, well-known general conservation laws will simply be mentioned. Laws specifically relevant to particle physics will be given some empirical justification but no attempt will be made to give a deeper theoretical underpinning.

7.3.1 Energy, Momentum and Angular Momentum Conservation

These three laws are well known and particle properties linked to them are mass and spin. These quantities are usually determined by making use of these conservation laws. The laws are universal and apply to all interactions.

7.3.2 Parity Conservation

As was discussed in section 6.5.3, parity conservation does not hold for the weak interaction. This was first proposed in 1956 by Lee and Yang in order to account for the fact that the K^+ can, *inter alia*, decay into $\pi^+ \pi^0$ or $\pi^+ \pi^+ \pi^-$. Since the former combination has even parity and the latter odd, this can only be understood if parity is not conserved. All the experimental evidence, however, supports the hypothesis that it is conserved in the strong and electromagnetic interactions. Further, it is possible to allocate an **intrinsic** parity to all hadrons. This is illustrated by considering the way in which the parity of the pion was first established.

The following capture process of a π^- by a deuteron (^{2}H, spin 1)

$$\pi^- + {}^2\text{H} \rightarrow \text{n} + \text{n}$$

takes place for pions at rest and so involves s-state (orbital angular momentum $l = 0$) capture. Thus, since the pion has $J = 0$, the total angular momentum after capture must be 1. The only state allowed by the Pauli exclusion principle for the two neutrons having total angular momentum 1 is the $^3\text{P}_1$ ($l = 1$) state which has odd ($(-1)^l$) parity. Since the deuteron has the same intrinsic parity as two neutrons (see below) it follows, if parity is conserved, that the π^- has intrinsic odd parity. Thus for the pion $J^P = 0^-$. Similar analyses of other strong interaction processes have established the parities of many mesons.

For baryons the allocation of an intrinsic parity is conventional since they cannot be created or annihilated in isolation as with the pion in the above example; they are always accompanied by an antiparticle which will have the opposite parity so the pair will always have odd intrinsic parity. By convention baryons are allocated even intrinsic parity and their antiparticles odd parity. On the contrary, for mesons, particle and antiparticle have the same parity.

In conclusion the reader is reminded that the photon has odd parity (section 6.4.2) and that parity is only conserved in the strong and electromagnetic interactions.

7.3.3 Charge Conjugation Invariance

Charge conjugation is the operation of converting a particle into its antiparticle and the charge conjugation operator is denoted by C. Thus

$$C\pi^+ \to \pi^-, \quad Cp \to \tilde{p}, \quad Ce^- \to e^+, \text{ etc.}$$

The strong and electromagnetic interactions are invariant under the charge conjugation operation. For example, the charge conjugate of the strong interaction process

$$p + \tilde{p} \to \pi^+ + \ldots$$

is

$$\tilde{p} + p \to \pi^- + \ldots$$

and experiment shows that, within experimental error, there is no difference in the π^+ and π^- energy spectra. Similarly, studies of the different electromagnetic decays of the η^0 (Table 7.1) have shown no evidence of charge conjugation non-invariance.

The weak interaction, on the contrary, is not invariant under charge conjugation. For example, it has been seen in section 6.5.3 that the v_e is left-handed (helicity $\lambda = -1$) and the $\tilde{v}_e$ is right-handed ($\lambda = +1$). But, operating with C on $v_e(\lambda = -1)$ gives

$$Cv_e(\lambda = -1) \to \tilde{v}_e(\lambda = -1)$$

leading to a particle (a left-handed antineutrino) which does not exist in the physical world as far as we know. However, the weak interaction and, for that matter, the strong and electromagnetic interactions, are invariant under the combined operations of charge conjugation (C) and reflection of axes (parity opertor P – section 2.6). Thus, for the v_e,

$$CPv_e(\lambda = -1) \to Cv_e(\lambda = +1) \to \tilde{v}_e(\lambda = +1)$$

leading to a physically allowed right-handed antineutrino. The first step follows since the mirror reflection of a left-handed entity is right-handed. In section 9.4 we shall see that, although CP reflection holds very widely, there is one important exception.

7.3.4 Charge Conservation

Charge conservation again is well known and has been verified with considerable accuracy. It applies to all interactions.

7.3.5 Lepton Conservation

One pair of leptons – (e^-, v_e) – has already been encountered in the discussion of β-decay. Two other pairs – (μ^-, v_μ) and (τ^-, v_τ) – with properties set out in Table 7.1 are now well established. They only experience the weak interaction and, in the case of charged leptons, the electromagnetic interaction. The absence of many energetically possible decay modes such as

$$\mu^- \to e^- + v_\mu + v_\mu$$

or

$$\tau^- \to \pi^- + v_e$$

or

$$\tau^- \to \mu^- + \mu^+ + \mu^-$$

or

$$n \to p + e^- + v_\mu$$

indicates clearly that the three pairs of leptons are different species. By allocating 'lepton numbers' L_e, L_μ and L_τ to the members of each species and requiring the total lepton number for each species to be conserved additionally in an allowed process, it is possible to account for all the experimental data. The following lepton numbers are conventionally assigned to each particle:

	e^-	v_e	μ^-	v_μ	τ^-	v_τ
L_e	+1	+1	0	0	0	0
L_μ	0	0	+1	+1	0	0
L_τ	0	0	0	0	+1	+1

The corresponding antiparticles take opposite values and all other particles have zero lepton number. Then in any allowed process involving leptons the sum of the lepton numbers for each species must remain constant. It is easy to check in the above examples that this is not the case (e.g. in the first example $L_\mu = 1$ on the left and $L_\mu = 2$ on the right, and so on). Lepton conservation applies to all interactions involving leptons (but see section 9.5).

7.3.6 Baryon Conservation

In the discussion of antiparticles in section 7.1 it was pointed out that, for fermions, particle and antiparticle can only be produced together and that they also annihilate each other. This is indicative of a conservation law similar to that applying to leptons. It is found, in fact, that the whole species of particles known as baryons are conserved in just the same way as each lepton species. Thus a baryon number $B = 1$ can be allocated to each baryon and a number $B = -1$ to each antibaryon with the requirement that the total baryon number is conserved in any allowed process. No elementary particle process has yet been observed violating this requirement. Thus this conservation law applies to all interactions involving baryons (but see section 9.5).

7.3.7 Isospin Conservation

Isospin (denoted by I and sometimes referred to as isobaric or isotopic spin) is a mathematical label given to all hadrons which reflects a basic symmetry found in the strong interaction. It has the same mathematical properties as ordinary spin but otherwise has nothing to do with physical angular momentum in space–time.

In section 3.3 it was mentioned that the internucleon potential is charge independent. That is to say it has the same form and strength between any two nucleons (nn, np or pp) provided that they are in the same quantum states. It is also to be noted that the masses of the neutron and proton only differ by 0.14%. Thus, as far as the strong interaction is concerned, neutron and proton can be regarded as different charge states of the same particle. These two states are analogous to the two different spin states of a nucleon ($J = \frac{1}{2}$, $J_z = \pm \frac{1}{2}$) and for this reason an **isospin** $I = \frac{1}{2}$ is allocated to a nucleon with the two values of $I_3 (= \pm \frac{1}{2})$ (it is conventional to use 1, 2 and 3 to indicate components in isospin space rather than x, y and z)

corresponding to proton and neutron as follows:

$$I_3 = +\tfrac{1}{2} \text{ signifies a proton}$$

$$I_3 = -\tfrac{1}{2} \text{ signifies a neutron}$$

With ordinary spin there is a spin operator J such that J^2 has eigenvalue $\tfrac{1}{2}(\tfrac{1}{2}+1)\hbar^2$ and J_z has eigenvalues $\pm\tfrac{1}{2}\hbar$. Similarly there is an isospin operator I such that I^2 has eigenvalue $\tfrac{1}{2}(\tfrac{1}{2}+1)$ and I_3 has eigenvalues $\pm\tfrac{1}{2}$.

Ordinary spin is, of course, an angular momentum and conservation of angular momentum stems from the invariance of a Hamiltonian under rotation of axes; it is a scalar quantity. In just the same way, the charge independence of nuclear forces implies that the internucleon potential must be a scalar in isospin space. Thus if it is to involve the isospins of the two nucleons it can only be in the form $I_1 \cdot I_2$ and isospin must be conserved.

Further, for two nucleons each having $I = \tfrac{1}{2}$, the total isospin can be 1 or 0. The different 3-components then correspond to the following configurations:

$$
\begin{array}{cccc}
I_3 = & +1 & 0 & -1 \\
I = 1 & \text{pp} & \text{np} & \text{nn} \\
I = 0 & & \text{np} &
\end{array}
$$

The two possible np states relate to the fact that neutron and proton are not identical particles (e.g. they have different electromagnetic properties) and so more states are allowed to them than to two protons or to two neutrons. In detail, the isospin structures of the $I = 1$, $I = 0$ neutron–proton states are

$$\psi(I = 1, I_3 = 0) = \frac{1}{\sqrt{2}}(\text{pn} + \text{np})$$

$$\psi(I = 0, I_3 = 0) = \frac{1}{\sqrt{2}}(\text{pn} - \text{np}) \qquad (7.2)$$

These states are completely equivalent in form to the structure of the $J = 1, 0$, $J_z = 0$ states for two spin $\tfrac{1}{2}$ particles.

The triplet, $I = 1$, state is reflected, for example, in the triplet of essentially identical states in the nuclei ^{14}O, ^{14}N, ^{14}C discussed briefly in section 6.5 and shown in Fig. 6.7. Here, the ground state of ^{14}N can be understood as a singlet ($I = 0$) isospin state; there is no equivalent state in ^{14}C and ^{14}O.

The allocation of isospin can be readily extended to the three pions. Again their strong interactions are observed to be essentially identical and their masses are very close (Table 7.1). But not there are three charge states and so the isospin must be $I = 1$ with

$$I_3 = +1 \text{ signifying the } \pi^+$$
$$I_3 = \ 0 \text{ signifying the } \pi^0$$
$$I_3 = -1 \text{ signifying the } \pi^-$$

Extending the previous discussion, any strong interaction Hamiltonian involving pions as well as nucleons must also be a scalar in isospin space and total isospin must be conserved in any process resulting from this interaction. This is dramatically confirmed in the following reaction:

$$^2\text{H} + {}^2\text{H} \rightarrow {}^4\text{He} + \pi^0$$

Both the deuteron (^{2}H) and ^{4}He have $I_3 = 0$ (they both have the same number of neutrons and protons) and, since there are no equivalent states such as ^{2}He (a 2p bound state) or ^{4}H (a 1p, 3n bound state) corresponding to $I_3 \neq 0$, they must both be isospin singlets with $I = 0$. Thus, since the pion has $I = 1$, it follows that isospin cannot be conserved in this process. It is found experimentally that its cross-section is at least two orders of magnitude less than the value expected if it was a strong interaction process, so confirming the conservation law. This is just one example of the many which strongly support the general hypothesis that isospin is conserved in strong interaction processes.

In Table 7.1 the isospins allocated to hadrons are given. For example

1. for *baryons*
   (a) $J = \frac{1}{2}$

Λ	$I = 0$
$\Sigma^-, \Sigma^0, \Sigma^+$	$I = 1$

   (b) $J = \frac{3}{2}$

$\Delta^-, \Delta^0, \Delta^+, \Delta^{++}$	$I = \frac{3}{2}$

2. for *mesons*
   (a) $J = 0$

K^+, K^0	$I = \frac{1}{2}$
$\eta' \ (958)$	$I = 0$

(b) $J = 1$

$$\rho^-(770), \rho^0(770), \rho^+(770) \quad I = 1$$

Turning now to the electromagnetic interaction it is easy to see that for nucleons and pions the electrical charge (Q) of a particle in units of e is related to the 3-component of isospin by the following relations

nucleons: $\quad Q = I_3 + \tfrac{1}{2}$

pions: $\qquad Q = I_3$

or for both sets of particles

$$Q = I_3 + B/2 \tag{7.3}$$

where B is the baryon number introduced previously. Clearly this expression does not work for most particles in Table 7.1 but we shall see below how it can be generalized for all hadrons. Even so, it is clear that charge is intimately related to I_3 and it follows that the electromagnetic interaction must depend on I_3. But I_3 is a vector in isospin space so that the electromagnetic interaction cannot be invariant under rotations in that space. It therefore does not conserve isospin. Similarly we shall see in Chapter 9 that weak interactions between hadrons also do not conserve isospin.

Leptons and the photon are not allocated an isospin; it is only meaningful to allocate it to those particles which experience the strong interaction – the hadrons. The corresponding antiparticles have the same isospin but opposite 3-component. Here it will be noted that in the case of antiprotons (for which $B = -1$) eq. (7.3) then leads correctly to $Q = -1$. Finally we restate the fact that isospin is only conserved in strong interaction processes.

7.3.8 Strangeness and hypercharge conservation

Although most experimental studies of elementary particle processes are now carried out using accelerators, in the immediate post World War II period the study of high energy cosmic rays was the only way of finding out about new particles. It was in this way that the pions and muons were identified. A particularly puzzling feature observed from 1947 onwards in cloud chamber experiments was the decay of some particles (originally referred to as V-particles) into V-shaped tracks. It is now known that what was being observed were decays of the following kinds:

$$\Lambda^0 \to p + \pi^-$$
$$K^0 \to \pi^+ + \pi^-$$

each decay having a mean life $\tau \simeq 10^{-10}$ s (Table 7.1). The puzzle lay in the fact that these particles were produced prolifically through what could only be a strong interaction process. However, if the Λ^0 was produced, for example, by the strong interaction process

$$p + \pi^- \to \Lambda^0$$

then, by detailed balance, τ for the inverse decay process should be characteristic of a strong interaction decay time, namely $\tau \sim 10^{-23}$ s! This contradiction led to the new particles' being referred to as 'strange'.

The contradiction was resolved in 1953 when strange particles were first produced in an accelerator. It was found that they were always produced in association with each other, a process referred to as **associated production**. A typical production process is

$$\pi^- + p \to \Lambda^0 + K^0$$

This phenomenon was codified by Gell-Mann and Nishijima who proposed that all hadrons had an additional quantum number S, referred to as **strangeness**, which had to be conserved additively in all strong interaction processes. For the particles in the above example the allocated values of S are

$$p(S=0), \quad \pi^-(S=0), \quad \Lambda^0(S=-1), \quad K^0(S=+1)$$

Clearly S is conserved additively in the associated production process but not in the decay process. The latter therefore was attributed to the weak interaction and its relatively long lifetime can be understood; the contradiction is resolved.

The values of S given to the various hadrons are displayed in Table 7.1 and it will be noted that some are high, e.g. $\Xi^{0,-}(S=-2)$, $\Omega^-(S=-3)$. This means that, taking the Ξ, it must be produced in association with particles having a total $S=+2$, for example two K mesons, each having $S=+1$. As is to be expected the strangeness of an antiparticle is simply the negative of the strangeness of the corresponding particle. So, although the $\pi^0(S=0)$ and its antiparticle are identical, the K^0 and $\tilde{K}^0$ are very different particles since they have $S=+1$ and $S=-1$ respectively.

It is now interesting to return to the expression $Q = I_3 + B/2$ (eq. (7.3)) relating the charge of non-strange hadrons to isospin and baryon number. This can now be simply modified so as to include strange hadrons by writing

$$Q = I_3 + \frac{B + S}{2} \tag{7.4}$$

and it is left to the reader to check that this assertion is true for all strange hadrons in Table 7.1. The above expression is frequently written in the form

$$Q = I_3 + Y/2 \tag{7.5}$$

where the new quantum number $Y = B + S$ is referred to as **hypercharge**.

Strangeness, then, and hypercharge are conserved in strong interaction processes but not always in weak interaction processes. For example, in the hadronic decays of the Λ^0 and K^0 given abot it can be seen that the strangeness changes by unity and it is generally observed that weak hadronic decays satisfy the selection rule $\Delta S = \pm 1$. For decays involving leptons and hadrons (semileptonic decays) S may change (e.g. $\Lambda^0 \rightarrow pe^-\bar{\nu}_e$) or may not change (e.g. $n \rightarrow pe^-\bar{\nu}_e$) and the implications of this will be discussed in Chapter 9. It should finally be noted that strangeness is conserved in electromagnetic processes.

Table 7.2 Conservation laws ('yes' implies conservation; 'no' implies non-conservation)

Quantity	Strong	Electromagnetic	Weak
energy	yes	yes	yes
momentum	yes	yes	yes
angular momentum	yes	yes	yes
parity	yes	yes	no
charge conjugation	yes	yes	no
charge	yes	yes	yes
lepton number	yes	yes	yes
baryon number	yes	yes	yes
isospin	yes	no	no
strangeness (hypercharge)	yes	yes	no

7.4 CONCLUSIONS

In this chapter some of the main intrinsic properties of particles and associated conservation laws have been described. The coverage has, however, not been exhaustive and further properties such as charm and bottomness will be dealt with in the next chapter in the context of more detailed discussions of the strong and weak interactions. This chapter concludes with a summary (Table 7.2) of the different conservation laws discussed and the interactions for which they hold.

8

The strong interaction

Since the formulation of the periodic table by Mendeleev in the last century and its explanation in terms of atomic structure during the early part of this century, there has been a constant search for ways of accounting for the regularities and symmetries observed in nature's building blocks. In the case of atoms a detailed understanding of the different families of similar atoms and their properties was achieved in terms of their electronic structures. Of course, each atom had a different nucleus but, in turn, the variety of nuclear species, as we have seen, can be understood in terms of their different nucleonic structures.

We have now reached the situation where the nucleons themselves and the pions which propagate the nuclear force are found to be members of extensive families of elementary particles. Since the late 1940s there has been a continuing attempt to understand the similarities and differences of particles in these families in terms of some simple underlying structure and in this chapter we outline the progress that has been made as far as strongly interacting particles (hadrons) are concerned.

8.1 NON-STRANGE HADRONS

Before the existence of strange particles had been established, the only known strongly interacting particles were the nucleons (n and p) and the pions (π^+, π^0, π^-). As we have seen the former belong to an isodoublet $(I = \frac{1}{2}, I_3 = \pm \frac{1}{2})$ and the latter to an isotriplet $(I = 1, I_3 = 0, \pm 1)$. One early attempt to understand the intrinsic properties of pions was made by Fermi and Yang in 1949. They pointed out that their properties could be understood if pions were regarded as bound states of a nucleon and an antinucleon, thus reducing the number of fundamental particles to two (n and p). Symbolically it was proposed that the three pions should have the following structures:

$$\pi^+ \equiv p\tilde{n}$$

$$\pi^0 \equiv \frac{1}{\sqrt{2}}(n\tilde{n} - p\tilde{p}) \tag{8.1}$$

$$\pi^- \equiv -n\tilde{p}$$

The following will be noted.

1. The lowest energy state of the nucleon–antinucleon pair will be an $S(l = 0)$ state and, further, to obtain the correct spin, $J = 0$ for the pion, the pair will need to be in a singlet state ($^1S_0$).
2. Since nucleon and antinucleon have opposite baryon numbers each pionic state must have $B = 0$ as required.
3. Since nucleon and antinucleon have opposite parity and are in a relative S-state (even parity), the overall parity is odd so that $J^P = 0^-$ as required.
4. Since I_3 has opposite values for nucleon and antinucleon it follows that the total I_3 has the correct values for the three pionic states (e.g. for the π^+, p has $I_3 = +\frac{1}{2}$ and $\tilde{n}$ also has $I_3 = +\frac{1}{2}$ so that the total value of I_3 is $+1$). It should also be recognized that the structures of the three isospin states are exactly equivalent in form to the triplet isospin states given in eq. (7.2). The differences in sign simply result from conventions used for the specification of antiparticle states.

There will, of course, be other states predicted. For example, there is the 0^- isosinglet state $(p\tilde{p} + n\tilde{n})/\sqrt{2}$ which could now be identified with the η^0, a particle not known by Fermi and Yang. Similarly, the corresponding set of 1^- particles formed when the nucleon–antinucleon pairs are in the *triplet* spin state could now be identified with the $\rho^{\pm,0}$ and the ω^0 (Table 7.1).

This scheme of things, since it takes no account of strangeness, is now known to be incorrect but it is a simple example of the way that thinking has developed over recent years. Basically, important intrinsic properties of a group of elementary particles were understood in terms of two fundamental constituents—the neutron and proton and their antiparticles. In particular the spins and isospins of the composite particles (in this case pions, etc.) result from the usual quantum addition of the spins $(J = \frac{1}{2})$ and isospins $(I = \frac{1}{2})$ of their constituents.

Although we shall not be using group theory it should be noted that the two spin (or isospin) $\frac{1}{2}$ states are the simplest basic states in what is known as the $SU(2)$ group (special unitary group in two dimensions) of transformations. The corresponding Pauli spin matrices are the simplest representation of this group which is intimately related to the group of rotations in ordinary or isospin space as the case may be. Here it will be remembered that it is invariance under such rotations that leads to the conservation of angular momentum and isospin. The simple results derived above for the isospin structure of pions and the η^0 in group theoretical symbolism are embodied in the following formal statement:

$$\mathbf{2} \otimes \tilde{\mathbf{2}} = \mathbf{3} \oplus \mathbf{1} \tag{8.2}$$

Here $\mathbf{2}$ and $\tilde{\mathbf{2}}$ signify doublet $(I = \frac{1}{2})$ isospin states for nucleon and antinucleon respectively, and the $\mathbf{3}$ and $\mathbf{1}$ signify the resultant triplet $(I = 1)$ and singlet $(I = 0)$ isospin states resulting from the combination of two such doublets. The circles surrounding the $\times$ and $+$ signs are to indicate that eq. (8.2) is not a simple arithmetical statement. As we shall see, this symbolism is a useful shorthand in the more extended considerations of the next section.

8.2 THE QUARK STRUCTURE OF HADRONS

To extend the ideas of the last section so as to include an understanding of strange particles the neutron and proton had to be abandoned as building blocks. In addition to an isospin doublet, necessary to understand the structure of pions and η_0, for example, an additional state with non-zero strangeness is clearly needed so that strange particles can be formed. Accordingly, in 1961 Gell-Mann and Ne'eman suggested that the basic symmetry underlying the different hadrons was $SU(3)$, the minimal extension of $SU(2)$ which can include strangeness.

Three years later Gell-Mann and Zweig went further and suggested that rather than being simply a mathematical basis for understanding the symmetries of hadrons, these three states actually corresponded to three fundamental spin $\frac{1}{2}$ particles from which the well-known particles were constituted. These were called **quarks** (denoted in general by q) by Gell-Mann, following James Joyce in *Finnegan's Wake*—'three quarks for Muster Mark...'. This suggestion was later supported by experiments at SLAC (section 5.3.1) in which high energy (tens of GeV) electrons undergo **deep inelastic scattering** by protons. By the latter phrase is meant inelastic scattering in which the electrons lose a large part of their energy. Experimentally it is found that such scattering results in the production of other hadrons and that the distribution of the inelastically scattered electrons can only be understood if there exist three point-like entities of spin $\frac{1}{2}$ in the proton with which the electrons interact. These entities are referred to as **partons** and can be identified with the proposed quarks. The discussion of hadron symmetries will now be continued in terms of quark structures.

In the $SU(3)$ scheme of things the mesons are built from quark and antiquark, (q$\bar{\text{q}}$), analogously to the structure suggested for mesons in the last section. Baryons, however, are built from three quarks, (qqq), and this at once implies that the baryon number of a quark is $B = \frac{1}{3}$. Two of the quarks would be expected to have strangeness $S = 0$ and to form an isodoublet, like the proton and neutron, so as to be able to form non-strange mesons as in the last section. Further, since, as has already been remarked, there is a baryon (Ω^-) with $S = -3$ it follows that the third quark must have $S = -1$ and that the Ω^- is built of three such quarks. This third quark, existing alone, must be an isospin singlet. However, because we now have three basic states ($SU(3)$ symmetry) two other spins in addition to isospin can be defined (referred to as U- and V-spin). With reference to them U- and V-doublets including the strange quark can be identified. The totality of these spins is called **unitary spin** and from the $SU(3)$ point of view there is complete symmetry between the three quarks.

In the light of these comments it is now straightforward to list the conjectured properties of each quark (Table 8.1). The names of the two non-strange quarks derive from their isospin components (up and down) and that of the third (strange) from its strangeness. Their charges, in units of e, are obtained by using the formula $Q = I_3 + Y/2$

Table 8.1 Properties of quarks

Symbol	Name	Q	I	I_3	B	S	Y
u	up	$+\frac{2}{3}e$	$\frac{1}{2}$	$+\frac{1}{2}$	$\frac{1}{3}$	0	$\frac{1}{3}$
d	down	$-\frac{1}{3}e$	$\frac{1}{2}$	$-\frac{1}{2}$	$\frac{1}{3}$	0	$\frac{1}{3}$
s	strange	$-\frac{1}{3}e$	0	0	$\frac{1}{3}$	-1	$-\frac{2}{3}$

given in eq. (7.5). In general the adjectives up, down and strange are referred to as quark **flavours** and we shall see in section 8.3 that three further flavours have to be introduced.

8.2.1 Hadron multiplets

In the case of mesons it is to be expected that the lowest energy states (i.e. mesons of lowest mass) will be q$\bar{\text{q}}$ combinations in an S-state. Because, like nucleons, the q and $\bar{\text{q}}$ have spin $\frac{1}{2}$ and opposite parity this will lead, as in section 8.1, to the formation of $J^P = 0^-$ and 1^- mesons. Further, just as a triplet of mesons and a singlet were formed by combining a nucleon and an antinucleon isospin doublet $(2 \otimes \bar{2} = 3 \oplus 1)$, so the combination of a quark and an antiquark triplet produces an octet and a singlet $(3 \otimes \bar{3} = 8 \oplus 1)$ of 0^- and also of 1^- mesons.

This means that, if the hypothesis is correct, it should be possible to identify an octet and a singlet of 0^- and 1^- mesons having similar strong interaction properties. This can be done and is demonstrated in Fig. 8.1 for the lowest mass 0^- and 1^- mesons. The diagrams presented are known as **weight diagrams** in which the horizontal axis gives the value of I_3 and the vertical axis the value of $Y (= B + S)$. Each particle is represented by a point on the diagram and is labelled with its symbol and generally with its quark–antiquark content. Two particles lie at the centre of the octet, for example the η^0 and the π^0 in the 0^- octet. In some cases the quark content is simple, e.g. $K^0 \equiv d\bar{s}$ and $\pi^- \equiv d\bar{u}$ whilst in others it is more complicated, e.g. $\pi^0 \equiv (d\tilde{d} - u\tilde{u})/\sqrt{2}$ (compare eq. (8.1)) and $\eta^0 \equiv (d\tilde{d} + u\tilde{u} - 2s\tilde{s})/\sqrt{6}$.

The quark content of the 1^- mesons is, of course, identical to that of the 0^- mesons; the difference between them lies in the fact that the quarks in the former are in a triplet spin state whilst in the latter they are in a singlet spin state. Having higher mass the members of the 1^- octet can be regarded as **excited states** of the 0^- mesons; hence the notation K^{*0}, etc.

Also given on the diagrams are the approximate masses of the different mesons. Here it will be noted that although they are the same for the different isospin multiplets they differ between multiplets. This is an example of symmetry breaking and shows that there is not complete $SU(3)$ symmetry between the three quarks.

Turning now to the baryons, they are hypothesized to have the structure qqq with antibaryons having the structure $\bar{\text{q}}\bar{\text{q}}\bar{\text{q}}$ and, for the lowest mass states, the three quarks are again expected to be in

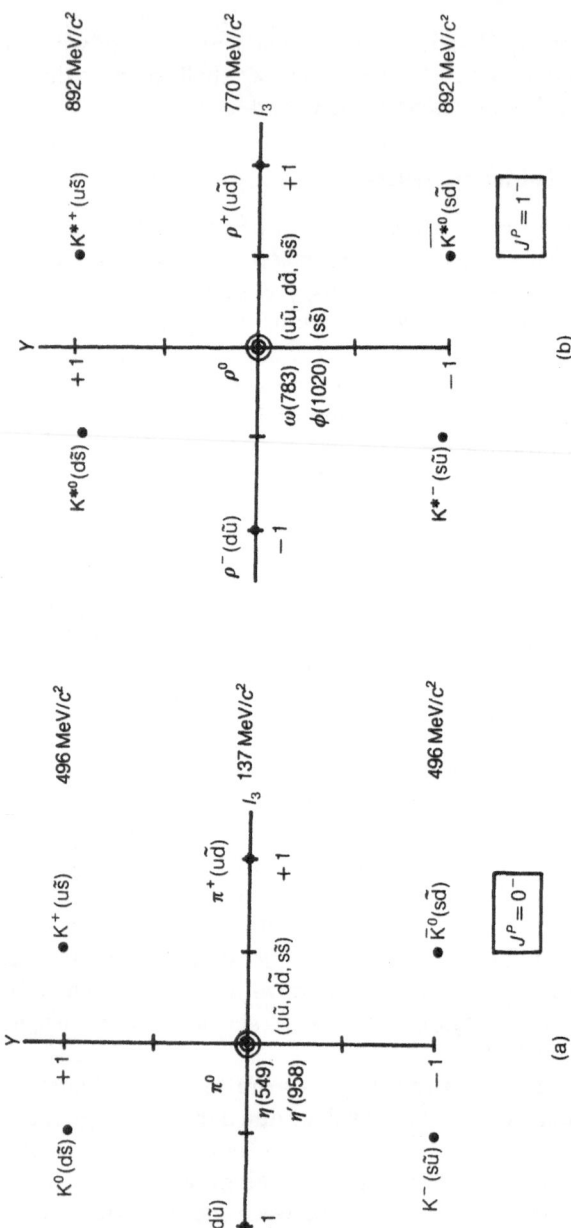

Fig. 8.1 Meson multiplets indicating quark content: (a) 0^- octet and singlet; (b) 1^- octet and singlet.

relative S-states. Thus the resultant hadrons should have $J^P = \frac{1}{2}^+$ or $\frac{3}{2}^+$. As far as multiplet structure is concerned we need to know the outcome of $3 \otimes 3 \otimes 3$. This can be shown to be $10 \otimes 8 \otimes 8 \otimes 1$ so that a decuplet and octets of baryons are to be expected. These are indeed found to exist in nature and the lowest $J^P = \frac{1}{2}^+$ octet and $\frac{3}{2}^+$ decuplet are shown in Fig. 8.2 on appropriate weight diagrams.

Again it is clear from the different masses of the various isospin multiplets that $SU(3)$ symmetry is by no means exact. Since the u and d quarks are members of an isospin doublet, they can be taken to have essentially equal masses. It then follows that since there are three of them in a nucleon (mass $\simeq 1$ GeV/c^2) they behave as though they have masses $m_u \simeq m_d \simeq \frac{1}{3}$GeV/$c^2 \simeq 330$ MeV/c^2. This is referred to as the **constituent** or **dynamical** mass and varies from hadron to hadron. It clearly embraces all the effects, as yet undiscussed, of the quark–quark interaction (equivalent to binding energy effects of nucleons in a nucleus).

Turning to the s quark, it is of particular interest to consider the different masses of the four isospin multiplets in the $\frac{3}{2}^+$ decuplet. The hadron masses in each successively heavier multiplet differ by around 140 MeV/c^2. But each multiplet differs by the replacement of a u or d quark by an s quark and so this mass difference can be understood if the s quark is heavier than a u or d quark by around 140 MeV/c^2, i.e. $m_s \simeq 470$ MeV/c^2. In the early 1960s when ideas about quarks were being developed the Ω^- had not been discovered and only nine members of the decuplet were known. Gell-Mann, however, noting the equal mass spacing of the multiplets was able to predict the existence and mass of the Ω^-. It was a tremendous triumph when in 1964 it was found in a bubble chamber experiment at Brookhaven (USA) with all the predicted properties. More detailed $SU(3)$ theory predicts other relationships between the masses of members of multiplets. For example in the $\frac{1}{2}^+$ baryon octet it is predicted that

$$3m_\Lambda + m_\Sigma = 2m_\Xi + 2m_p$$

and the agreement with experiment is remarkable (left-hand side, $\simeq 4540$ MeV/c^2; right-hand side, $\simeq 4520$ MeV/c^2).

Many other multiplets have been identified and the quark description of hadrons is now well established. However, there are some particles which are not accommodated within the quark structure given so far and additional quarks have had to be introduced. These will be discussed in section 8.3, but first we look a little more carefully into the quark structure of the multiplets so far encountered.

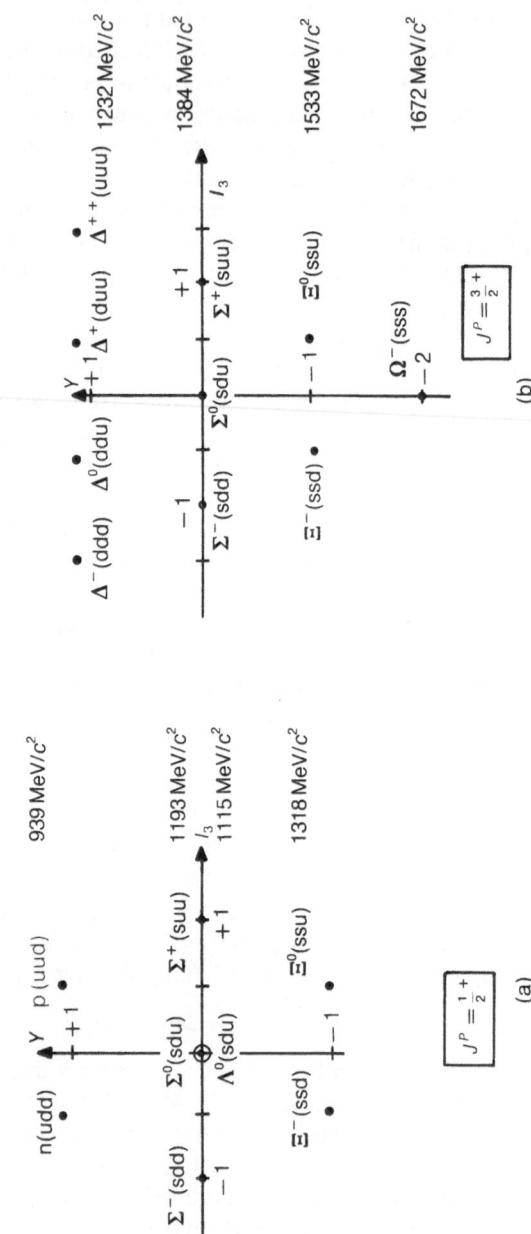

Fig. 8.2 Baryon multiplets indicating quark content: (a) $\frac{1}{2}^{+}$ octet; (b) $\frac{3}{2}^{+}$ decuplet.

8.2.2 Antisymmetry and Colour

In the discussion so far, the quark content of different hadrons has been identified, their parities have been attributed to the component quarks being in relative S-states and their spins result from the vector addition of the component quark spins. A further detail must now be tackled, namely the antisymmetry of the intrinsic particle wave function required by the Pauli exclusion principle under the exchange of identical quarks. The need for this manifests itself most strikingly in the cases of the Δ^- (ddd), the Δ^{++} (uuu) and the Ω^- (sss) each of which contains three identical quarks.

The spatial part of the wave function (an S-state) is clearly symmetrical depending only on the magnitude of the spatial separation between quarks. Further, since each particle has $J = \frac{3}{2}$, the spin part of the wave function is also symmetrical (for example, in the $J_z = \frac{3}{2}$ state each quark has $s_z = \frac{1}{2}$). Thus, as it stands, the overall wave function for these particles is symmetrical under the exchange of quarks. This, of course, is not allowed and therefore a further attribute, like charge, must be allocated to quarks so that a wave function having overall antisymmetry can be constructed.

A way in which this might be done was proposed in 1964 by Greenberg who suggested that this attribute should be called **colour**, with an associated colour charge, and that there should be three varieties – **red**, **yellow** and **blue**. Three are clearly necessary since there are three possible quark–quark interchanges in a hadron; or, in exclusion principle language, each of the three identical quarks must be in a different colour state since their space and spin states are identical. Colour is rather like an extended version of electric charge except that it comes in three varieties. In due course (section 8.4) we shall see that this analogy is far from trivial and that just as charge couples the carrier of the electromagnetic interaction (the photon) to a particle (section 3.4) so colour couples the carrier of the interquark interaction (called the **gluon**) to a quark.

Introducing colour in this way opens the door to a much wider variety of elementary particles, distinguished by different overall colour, than is, in fact, observed. No 'coloured' elementary particles have so far (1990) been detected. This suggests that all hadrons are colour **singlets** or, in colour language, noting the effect of mixing the three primary colours, are 'white'. Note here that the mesons, having the essential structure $q\bar{q}$, will also be colour singlets since the colour of the quark will be cancelled by that of the antiquark.

The three colour states of a quark are now regarded as a three-dimensional representation of an $SU(3)$ **colour group**. Here we note from earlier discussion of $SU(3)$ symmetry in this section that combining three quarks having three different colour states does indeed lead, *inter alia*, to a singlet state

$$3 \otimes 3 \otimes 3 = 1 \oplus 8 \oplus 8 \oplus 10$$

as does the combination of quark and antiquark

$$3 \otimes \tilde{3} = 1 \oplus 8$$

Other combinations of quark and antiquark (e.g. qq$\tilde{q}$, qq,...) can be shown not to lead to a colour singlet. Thus the requirement that all hadrons are colour singlets restricts the combinations of quarks observed in nature just to the two structures qqq and q$\tilde{q}$. Finally, it is re-emphasized that the $SU(3)$ colour symmetry is only coincidentally the same as the flavour (u, d, s) $SU(3)$ symmetry discussed earlier in this section. Both symmetries happen to be based on the $SU(3)$ group, but are physically quite different. The flavour symmetry, because of the different quark masses, is, as already stated, only approximate whereas colour $SU(3)$ symmetry is believed to be exact.

Knowing the symmetry of the quark wave function it is now possible to obtain information about the magnetic moments of baryons; in particular, those of the neutron and proton. In the case of the proton the quark content is uud. Since the proton wave function must be antisymmetric under the exchange of the two u quarks, it follows that they must be in the symmetric $S = 1$ state where S is the total spin of the two u quarks. Antisymmetry then follows since the space part of the wave function is also symmetric but the colour part is, as discussed above, antisymmetric.

The quark magnetic moment operators are taken to have the usual general form (section 2.7.1), namely

$$\mu_u = g \frac{Q_u}{2m} s_u \quad \mu_d = g \frac{Q_d}{2m} s_d \tag{8.3}$$

where g is the quark g-factor, m is the (common) constituent mass of the u and d quarks and $Q_{u,d}$ are their charges. Substituting for the latter from Table 8.1, the magnetic moment operators can be written more simply

$$\mu_u = \lambda_u \mu s_u \quad \mu_d = \lambda_d \mu s_d \tag{8.4}$$

where $\mu = ge/2m$, $\lambda_u = 2/3$ and $\lambda_d = -1/3$.

The magnetic moment operator for the $S = 1$ combination of two u quarks is obtained by adding the magnetic moments of the two quarks, giving

$$\mu_{uu} = \lambda_u \mu S \tag{8.5}$$

where $S(= s_u(1) + s_u(2))$ is the total spin operator for the two u quarks (labelled 1 and 2). Thus the proton magnetic moment operator (μ_p), which is simply the sum of the contributions from the two u quarks and the d quark, is given by

$$\mu_p = g_p \frac{e}{2m_p} J = (\lambda_u S + \lambda_d s_d)\mu \tag{8.6}$$

where μ_p is initially written in the form given in eq. (2.16) with s_p replaced by J (the symbolism now used for a particle spin operator). J, of course, is simply the sum of S and s_d, i.e.

$$J = S + s_d$$

or

$$s_d = J - S \tag{8.7a}$$

or

$$S = J - s_d \tag{8.7b}$$

To calculate μ_p, the same procedure can be used as in section 4.2.3 for calculating magnetic moments on the shell model. Equations (8.6) and (8.7a, b) play exactly the same role as eqs (4.8) and (4.10a, b) in that calculation. Thus eq. (8.6) is scalar multiplied by J and the resultant terms $S \cdot J$ and $s_d \cdot J$ replaced by the values obtained from squaring eqs (8.7a, b) respectively. Replacing J^2, S^2 and s_d^2 in the resultant expression by their respective eigenvalues $-3\hbar^2/4$, $2\hbar^2$ and $3\hbar^2/4$ – gives finally

$$\frac{eg_p}{2m_p} = (4\lambda_u - \lambda_d)\frac{\mu}{3} = \mu \tag{8.8}$$

An exactly analogous calculation can be carried out for the neutron magnetic moment μ_n. We can go from a proton to a neutron by simply interchanging u and d quarks, so that

$$\frac{eg_n}{2m_p} = (4\lambda_d - \lambda_u)\frac{\mu}{3} = -\frac{2}{3}\mu \tag{8.9}$$

The ratio of the nucleon magnetic moments is therefore predicted to be $\mu_n/\mu_p = -2/3$. This is to be compared with the experimental result (section 2.7.1 gives values of g_n and g_p) $\mu_n/\mu_p = -0.685$. The agreement is clearly good.

Magnetic moments of the other baryons can similarly be calculated in terms of, say, μ_p and insofar as these moments have been measured there is fair to good agreement between theory and experiment. This agreement confirms our confidence in the basic quark model.

8.3 ADDITIONAL QUARKS

The hadrons so far discussed have been structured from three basic quarks – u, d and s – each of which can have three different colours. In the early 1970s, however, it became clear that there existed hadrons whose properties could not be accounted for in terms of these quarks. Although there were earlier indications of this, the first definitive evidence came in 1974 from two different sources in the USA.

At the Brookhaven National Laboratory (BNL), the bombardment of Be by 30 GeV protons led to the production of e^-e^+ pairs of definite energy (3097 MeV) in the centre-of-mass system suggesting that they resulted from the decay of a particle or resonance, referred to as the J meson, having that mass energy. Experiments at the Stanford Linear Accelerator Center (SLAC), involving colliding beams of very high energy ($\simeq 2$ GeV) electrons and positrons, revealed a resonance (referred to as the ψ meson) at the same energy. Clearly both experiments had detected the same particle which is now referred to as the J/ψ meson. The striking feature of this meson was that its width was very small ($\simeq 0.06$ MeV) compared with the width ($\simeq 100$ MeV) expected for a strong interaction decay process. For a particle whose mass was greater than 3 GeV/c^2, the natural expectation was that it would decay via the strong interaction into some combination of lower mass hadrons of which there are many candidates.

The explanation suggested was that a new quark flavour was involved which was called **charm** (denoted by c) and that the J/ψ had the structure c$\bar{c}$. The c quark is taken to have charm $\mathfrak{C} = +1$, all other quarks having charm 0. It also has $Q = +2e/3$, $S = I = 0$. The process observed at the BNL was then interpreted as due to the formation of a J/ψ in the initial bombardment which subsequently transformed via a virtual photon into an e^-–e^+ pair (Fig. 8.3a). The SLAC result involved the inverse process (Fig. 8.3b). Since the proposed process takes place via a photon ($J^P = 1^-$), angular momentum and parity conservation require that the J/ψ is also a 1^- particle.

Like strangeness, charm was postulated to be conserved additively

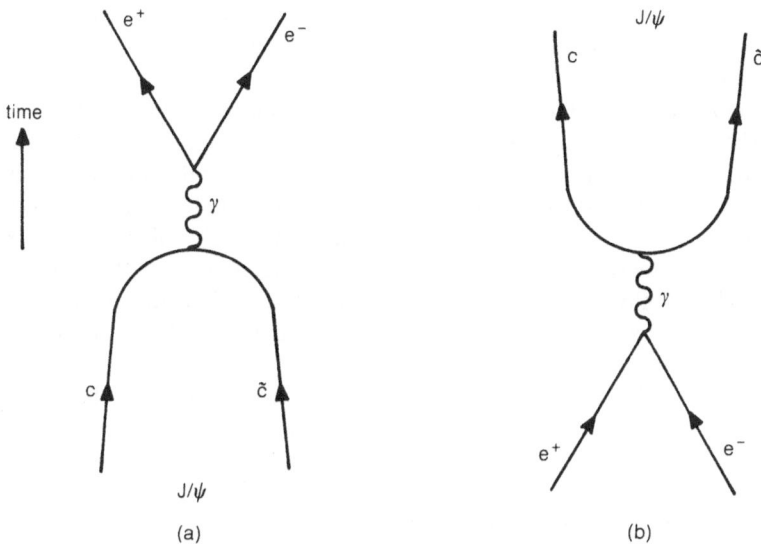

Fig. 8.3 (a) Decay and (b) creation of a J/ψ meson.

in strong and electromagnetic interactions and at first sight it is therefore not clear why the J/ψ, which has charm 0 (c and c̃ have charm $+1$ and -1 respectively), does not decay via the strong interaction into other lighter mass hadrons. The reason for this is that processes in which the flavour of the initial quarks changes are inhibited. In this case there are initially c and c̃ quarks but there would be no such quarks in the product hadrons. This is known as the Zweig rule and is also seen to operate in the case of the ϕ($J^P = 1^-$), which has the quark structure sš̃ (Table 7.1 and Fig. 8.1b). This decays readily into a K^+ (uš̃) and a K^- (sũ), which contain s quarks, but not so readily into lighter π-mesons which contain no s quarks. Without Zweig inhibition the 3π decay would be expected to be at least two orders of magnitude stronger.

In the case of the J/ψ, uninhibited hadronic decay could take place if, for example, charmed mesons containing a single c or c̃ and of sufficiently low mass existed. Charmed mesons of this kind have now been identified, but the lightest (Table 7.1) – the D^+ (cd̃) and D^0(cũ) – have masses too large ($\simeq 1865 \text{ MeV}/c^2$) for two of them to be emitted in the decay of the J/ψ. However, excited states of the J/ψ (designated ψ) have been detected as higher energy resonances in $e^+ - e^-$ collider experiments and the dominant decay mode of the ψ (3770), for example, is indeed into a D and a $\tilde{\text{D}}$.

Several such excited states of the J/ψ meson have now been identified in which the c and $\tilde{\text{c}}$ quarks are in relative S, P and D states. These states are analogous to the excited states of the e^+e^- system, known as **positronium**. The c$\tilde{\text{c}}$ system is accordingly referred to as **charmonium** and the similarity of the excited state level structures of positronium and charmonium, although on vastly different energy scales, suggests that the interquark potential has some of the features of the Coulomb potential. This point will be taken up in the next section. Charmed baryons have also been identified, for example the Λ_c^+ (udc) (Table 7.1).

Clearly the constituent mass of a c quark must be greater than the s quark since the masses of particles containing them are much heavier than uncharmed hadrons. By fitting the charmonium spectrum and from the masses of the D mesons the c quark constituent mass is estimated to be $\simeq 1580\,\text{MeV}/c^2$.

In a similar fashion to the discovery of charm at the BNL a further quark (**bottom** or **beauty**; denoted by b and with 'bottomness' $\mathfrak{B} = -1$) was identified in 1977 when resonances in the region of 10 GeV with very narrow decay widths were observed decaying to $\mu^+\mu^-$ pairs. These were attributed to different states of mesons (denoted by Υ; Table 7.1) having the quark structure b$\tilde{\text{b}}$. The existence of these different states of the Υ was confirmed later in e^+e^- collider experiments where it was also found that Υ states of higher energy could decay hadronically into mesons (B^+, B^0; Table 7.1) containing a b or $\tilde{\text{b}}$ quark. Again, by considering the masses of the different Υ states and of the B mesons it is estimated that the mass of the b quark is $\simeq 4580\,\text{MeV}/c^2$. It has charge $-e/3$ and $S = I = 0$. Like charm all other hadrons have zero bottomness and bottomness is conserved additively in strong and electromagnetic processes.

A further, and probably final, quark (**top** or **truth**; denoted by t and with 'topness' $\mathfrak{T} = +1$) with charge $+2e/3$ has also been conjectured for symmetry reasons. Six basic leptons (e, μ, τ and their corresponding neutrinos) have been identified (Table 7.1) and symmetrical theories require the existence of six corresponding quarks. However, it has not yet (1990) been detected. This alone sets a lower limit on its constituent mass in the region of $90\,\text{GeV}/c^2$, but there are compelling theoretical reasons for believing that it must be lighter than about $250\,\text{GeV}/c^2$.

In Table 8.2 the additive quantum numbers of all the quarks are

Table 8.2 Additive quantum numbers of quarks

Quark type (flavour)	u	d	s	c	b	t
baryon number (B)	$\frac{1}{3}$	$\frac{1}{3}$	$\frac{1}{3}$	$\frac{1}{3}$	$\frac{1}{3}$	$\frac{1}{3}$
electric charge (Q)	$+\frac{2}{3}$	$-\frac{1}{3}$	$-\frac{1}{3}$	$+\frac{2}{3}$	$-\frac{1}{3}$	$+\frac{2}{3}$
Isospin (I_3)	$+\frac{1}{2}$	$-\frac{1}{2}$	0	0	0	0
strangeness (S)	0	0	-1	0	0	0
charm $\mathfrak{C}$	0	0	0	$+1$	0	0
bottomness $\mathfrak{B}$	0	0	0	0	-1	0
topness $\mathfrak{J}$	0	0	0	0	0	$+1$

gathered together for convenience. Note that the charges of all quarks, in particular, and all hadrons, in general, are given correctly by extending the relationship given in eq. (7.4) to

$$Q = I_3 + \tfrac{1}{2}(B + S + \mathfrak{C} + \mathfrak{B} + \mathfrak{J}) \qquad (8.10)$$

8.4 THE QUARK–QUARK INTERACTION

The discussion in the previous sections has been carried out with the implicit assumption that there is an attractive interaction between quarks and antiquarks binding them into the various structures hypothesized for the different hadrons. There have already been some indications of its properties. First, the correspondence between charmonium and positronium states suggests that it has some similarities to the Coulomb potential between oppositely charged particles although, of course, much stronger. Second, the fact that $J = 3/2$ baryon states have much higher energy ($\simeq 300\,\text{MeV}$) than $J = 1/2$ states, having the same quark content, suggests that the interaction is strongly spin dependent. Similarly, the same sort of energy difference also arises between 0^- and 1^- mesons. Third, and most intriguing, is the fact that no isolated quarks have ever been detected in spite of many very thorough searches and their easily recognizable signature of one-third and two-thirds integral charges. This implies that the interquark potential must have some feature which prevents the forcible disintegration of a hadron into its component quarks.

With this in mind, consider first the form of the electromagnetic potential, resulting from photon exchange (Fig. 3.6), between, say, an electron and a proton as calculated by quantum electrodynamics

(QED). This has three main components

$$V_{em} = V_C + V_{LS} + V_{SS} \tag{8.11}$$

where V_C is the usual Coulomb interaction, V_{LS} is a spin–orbit term resulting from the interaction of the electron magnetic moment with the magnetic field it experiences whilst orbiting the proton and from relativistic effects, and V_{SS} is a spin–spin potential due to the interaction of the electron and proton magnetic moments. These last two terms are responsible for fine structure and hyperfine structure effects in the hydrogen atom.

V_C is well known and given by

$$V_C = -\frac{e^2}{4\pi\varepsilon_0 r} = -\frac{e^2}{4\pi\varepsilon_0 \hbar c}\frac{\hbar c}{r} = -\alpha\frac{\hbar c}{r} \tag{8.12}$$

where α is the fine structure constant ($\simeq 1/137$). We shall not be concerned with the details of V_{LS} and V_{SS} but note for future reference that V_{SS} has the order of magnitude

$$V_{SS} \sim \frac{1}{4\pi\varepsilon_0 c^2}\frac{\mu_p\mu_e}{r^3} \tag{8.13}$$

where μ_p ($\sim e\hbar/2m_p$) and μ_e ($\sim e\hbar/2m_e$) are the magnetic moments of the proton and electron respectively. Substituting for μ_p and μ_e then gives, after a little rearrangement and neglecting numerical factors,

$$V_{SS} \sim \alpha\frac{\hbar^3}{m_p m_e c r^3} \tag{8.14}$$

Turning now to the corresponding interaction between two quarks, this is attributed to the exchange of massless particles called **gluons** which couple to the colour charge of a quark just as the massless photon couples to electric charge (Fig. 8.4). Of course the situation is

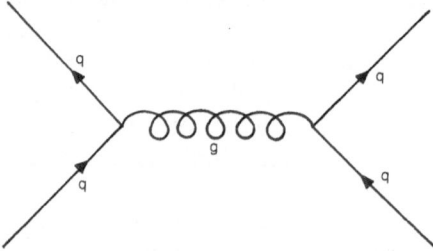

Fig. 8.4 Quark–quark interaction propagated by gluon (g) exchange.

much more complicated since there are three colour charges. The theory describing this interaction is, like QED, a gauge theory invariant under certain gauge transformations. It is known as **quantum chromodynamics** (QCD) and reflects the $SU(3)$ colour symmetry of the quarks referred to in section 8.2.2. This symmetry requires the existence of an octet of gluons as against a single photon in QED. Further, unlike the photon which carries no electric charge, the gluons actually carry colour charge and this has a profound effect on the nature of the interquark potential. What is found is that this (strong) potential has the form

$$V_S \sim -\alpha_s \frac{\hbar c}{r} + V_{sSS} + \ldots + `\lambda r' \tag{8.15}$$

where α_s ($\simeq 1$, section 3.4) measures the strength of the strong interaction and is the QCD analogue of the fine structure constant. It has the same form as α, namely $\alpha_s = g_s^2/\hbar c$ where g_s is now a measure of the colour charge. Actually, detailed theory shows that the value of α_s depends on the separation of the interacting quarks. Thus, although α_s is of the order 100α at separations of 10^{-15} m – indicated by the fact that cross-sections for hadronic processes at energies around 1 GeV (for which the de Broglie wavelength $\lambda = h/p \sim 10^{-15}$ m) are around 10^4 times larger than for electromagnetic processes – as the separation decreases $\alpha_s \to 0$. This is referred to as **asymptotic freedom**. QCD is the underlying theory which describes the strong interactions of hadrons, and the strong internucleon interaction is simply a manifestation of QCD. Thus, the main features of the internucleon potential, previously attributed to meson exchange (section 3.4), should now be understood in terms of the exchange of virtual $q\bar{q}$ pairs.

The first term is of Coulomb form but two orders of magnitude larger ($\alpha_s \simeq 100\alpha$) and so ties in with the expectations at the beginning of this section.

V_{sSS} is the analogue of V_{SS} and has an equivalent form to that given in eq.(8.14), namely

$$V_{sSS} \sim \alpha_s \frac{\hbar^3}{m_q^2 c r^3} \tag{8.16}$$

where m_q is now the constituent quark mass. The order of magnitude of this term can be obtained by taking r to be $\simeq 10^{-15}$ m, the approximate size of, for example, a nucleon. Using the constituent mass of u, d quarks ($330 \, \text{MeV}/c^2$) gives $V_{sSS} \sim 100 \, \text{MeV}$. This is the correct order of magnitude to account for the strong spin dependence of the interquark potential referred to earlier.

There are additional terms including a spin–orbit potential but, since the quarks in the hadrons we have considered are all in S-states, it makes no contribution.

The final term, 'λr', is fundamentally different from all the others and has no QED analogue. It is known as the **confinement** term and is in quotes since the linear dependence on r is a considerable simplification of a much more complex reality. Nevertheless, this phenomenological representation is used for more detailed calculations of hadronic structures, for example in accounting for charmonium energy levels. This term arises because the gluons carry colour charge. As a result, the lines of colour force between two quarks are attracted together to form a narrow tube of force (Fig. 8.5a). In the electrical case, on the contrary, lines of force spread out to infinity (Fig. 8.5b). Assuming uniform energy density throughout the tube, it then follows that the potential energy is proportional to r. It is as though the quarks are attached to each other by a piece of elastic!

One important implication of this confinement term is that it is impossible to remove a quark from a hadron to infinity, i.e. to obtain an isolated quark. Consider an attempt to remove a quark from a meson (q$\bar{\text{q}}$) by bombarding the meson with some very high energy

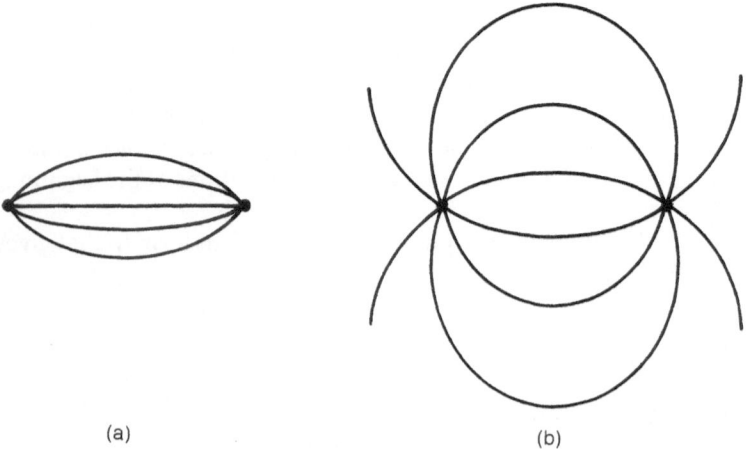

(a)   (b)

Fig. 8.5 Lines of force between (a) two colour charges and (b) two electrical charges.

particle. As the q and q̃ separate, the flux tube connecting them extends so that the energy builds up (linearly) and, at some point, it becomes energetically possible to form a qq̃ pair. The q and q̃ in this pair can then join respectively with the initial q̃ and q to form two mesons and the tube breaks. Multiple meson production can happen at higher energies, and similar processes can take place in attempting to disintegrate a baryon. The upshot is that, however much energy is put into trying to disintegrate a hadron, the quarks and gluons remain confined and all that results is the production of further hadrons.

Physical evidence for the existence of gluons comes from a number of sources. Most striking is in the formation of hadronic 'jets' following high energy (tens of GeV energy in the centre-of-mass system) e^+e^- collider experiments. The process is illustrated in Fig. 8.6a and is of the form

$$e^+ + e^- \rightarrow \gamma \rightarrow q + \tilde{q}$$

where the quark and antiquark fragment into collimated hadron jets since they cannot remain isolated. Since in a collider experiment the centre of mass is at rest, the two jets are emitted back to back. These jets are fairly narrow since the q and q̃ have little transverse

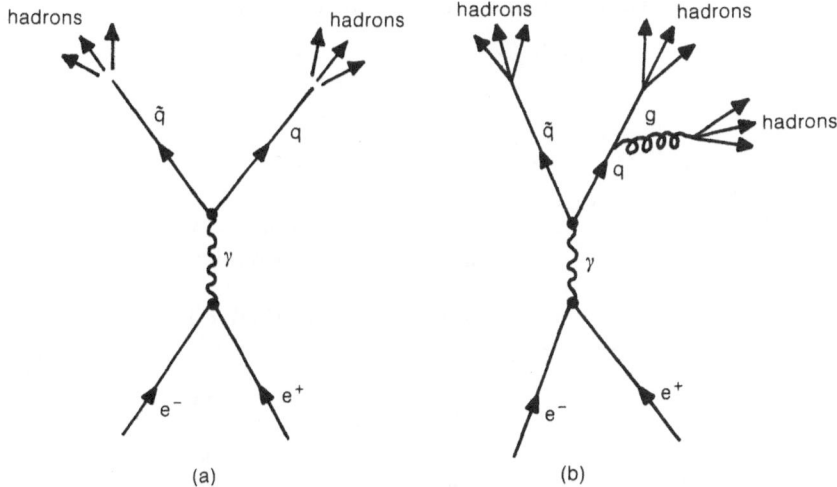

Fig. 8.6 (a) Hadron jets from e^+e^- collisions and (b) additional jet at higher energy due to gluon emission.

momentum relative to the e^+e^- beams. However, as the energy is increased, a third jet starts to appear and this can be attributed to the emission of a gluon by the q or q̄ (Fig. 8.6b) and is strong experimental evidence for the existence of gluons.

Finally, we return to the mass of quarks which have so far only been discussed including all dynamical effects (the **constituent** mass). The more fundamental mass to consider is what is known as the **current** mass which is the mass free of dynamical effects and is the mass which occurs in the basic theory of QCD. Suffice it to say here that these masses are somewhat smaller (particularly for the u and d quarks) than the constituent masses (approximate values are given in Table 8.3 and Kenyon (1987)).

It will be noted that the current mass of the d quark is greater than that of the u quark. This is to account for the fact that the neutron (udd) is heavier than the proton (uud). Furthermore, the current masses of the u and d quarks are much lighter than the nucleon mass ($\simeq 1\,\text{GeV}/c^2$). This means that the quark structure of a nucleon is highly relativistic since, to confine a quark to a region of nucleon size ($R \simeq 10^{-15}\,\text{m}$) implies by the uncertainty principle that its momentum has the order of magnitude

$$p_x \simeq p_y \simeq p_z \simeq \hbar/R \simeq 1 \times 10^{-19}\,\text{kg m s}^{-1} \simeq 200\,\text{MeV}/c$$

Such a momentum is relativistic for particles of mass less than $10\,\text{MeV}/c^2$ and so the energy can be calculated using $E = pc$, i.e.

$$E = (p_x^2 + p_y^2 + p_z^2)^{1/2}c \simeq 340\,\text{MeV}$$

Table 8.3 Quark masses

Quark	Current mass (MeV/c^2)	Constituent mass (MeV/c^2)
d	8.5	330
u	5.5	330
s	150	470
c	1250	1580
b	4250	4580
t	> 90 000	> 90 000

This is the energy of a single quark corresponding to a constituent mass of 340 MeV/c^2, in good agreement with our earlier estimate.

In conclusion it must be stressed that the discussion in this section is a highly simplified description of a very complicated theory. Nevertheless, it should give some idea of present-day thinking about the nature of the strong interaction.

9

The weak interaction and unification

9.1 INTRODUCTION

Looking through the list of elementary particles given in Table 7.1 it can be seen that lifetimes fall into three categories. First, there are those decays characterized by a 'width' (Γ) rather than a lifetime. These widths are measured in MeV and correspond to lifetimes of the order $\tau(\simeq \hbar/\Gamma) \simeq 10^{-21}\,\text{s}{-}10^{-23}\,\text{s}$. Decays of this kind are strong interaction processes in which all conservation laws are satisfied, in particular conservation of isospin, hypercharge (strangeness) and parity. Second, there are a few processes (e.g. $\pi^0 \rightarrow 2\gamma$, $\Sigma^0 \rightarrow \Lambda\gamma$) with somewhat longer lifetimes which are electromagnetic in origin; they conserve hypercharge and parity but not isospin. Third, there are many decays with lifetimes greater than $\simeq 10^{-13}\,\text{s}$. These decays, of which nuclear β-decay discussed in section 6.5 *et seq.* is an important example, are attributed to the weak interaction and, in particular, do not conserve parity. It is with these decay processes that we now concern ourselves. They can be divided into three varieties depending on the extent to which leptons are involved. These varieties are referred to as (a) leptonic, (b) semileptonic and (c) non-leptonic decay.

9.1.1 Leptonic Decays

The leptonic decays are the decays of the μ^- and the μ^+, namely

$$\mu^- \rightarrow \nu_\mu + e^- + \tilde{\nu}_e$$
$$\mu^+ \rightarrow \tilde{\nu}_\mu + e^+ + \nu_e$$

Note that these processes only involve leptons and that they conserve the lepton numbers L_e and L_μ (section 7.3).

It will be recognized that they have a similar form to the prototype β-decay processes (section 6.5)

$$n \rightarrow p + e^- + \tilde{v}_e$$
$$p \rightarrow n + e^+ + v_e$$

and, as in them, the v_e and v_μ both have helicity $\lambda = -1$ – they are both left-handed. The two processes can be analysed on an exactly equivalent basis to β-decay except that, because of the large energy release ($\simeq 105$ MeV), all particles must be treated relativistically. In particular, the processes can be described in terms of two coupling constants $G_{\mu V}$ and $G_{\mu A}$ equivalent to the G_V and G_A of β-decay (section 6.5). Comparing experimental data on the decays with theory it is found that

$$G_{\mu V} = -G_{\mu A} = G_\mu = 1.435 \times 10^{-62} \, \text{J m}^3$$

or, alternatively,

$$G_\mu/(\hbar c)^3 = 1.166 \times 10^{-5} \, \text{GeV}^{-2} \qquad (9.1)$$

This latter expression is in the units currently used in elementary particle physics. There are, of course, small experimental errors ($\sim 0.01\%$) in these results, but these are not included since they do not affect the subsequent discussion.

9.1.2 Semileptonic Decays

These are decays involving both hadrons and leptons. Typical examples for the cases $\Delta S = 0$ and $\Delta S = \pm 1$ are:

1. for $\Delta S = 0$

$$n \rightarrow p + e^- + \tilde{v}_e$$
$$\pi^- \rightarrow \mu^- + \tilde{v}_\mu$$
$$\rightarrow e^- + \tilde{v}_e$$
$$\mu^- + {}^A_Z X \rightarrow {}^{~~A}_{Z-1} X + v_\mu$$

2. for $\Delta S = \pm 1$

$$K^+ \rightarrow \pi^0 + e^+ + v_e$$
$$\Lambda^0 \rightarrow p + e^- + \tilde{v}_e$$

The last $\Delta S = 0$ process is not strictly a decay process and involves the capture of a μ^- by a nucleus; it is referred to as **muon capture** and is similar to electron capture (section 6.5).

The strength of these processes can again be characterized by coupling constants. Indeed, we have already seen (section 6.5) that for the first $\Delta S = 0$ process (β-decay)

$$G_V(\Delta S = 0) = 1.397 \times 10^{-62} \, \mathrm{J \, m^3}$$

or

$$G_V(\Delta S = 0)/(\hbar c)^3 = 1.135 \times 10^{-5} \, \mathrm{GeV^{-2}} \tag{9.2}$$

Again experimental error ($\sim 0.1\%$) is omitted. It will be noted that, although close to G_μ above, G_V does not have exactly the same value; thus

$$G_V(\Delta S = 0)/G_\mu = 0.973 \tag{9.3}$$

Further, we have seen in section 6.5 that $G_A \simeq -1.27 \, G_V$.

Turning to the $\Delta S = \pm 1$ processes, it is found experimentally that they are significantly weaker than $\Delta S = 0$ processes with

$$G_V(\Delta S = \pm 1) = 0.255 \times 10^{-5} \, \mathrm{GeV^{-2}} \tag{9.4}$$

so that

$$G_V(\Delta S = \pm 1)/G_\mu = 0.219 \tag{9.5}$$

9.1.3 Non-Leptonic Decay

These processes do not conserve parity, isospin or strangeness and are characterized by the selection rules $\Delta S = \pm 1$, $\Delta I = \pm 1/2$. Typical examples are

$$\Lambda^0 \rightarrow p + \pi^-$$
$$\Sigma^+ \rightarrow p + \pi^0$$
$$\Xi^- \rightarrow \Lambda^0 + \pi^-$$
$$\Omega^- \rightarrow \Lambda^0 + K^-$$
$$K^+ \rightarrow \pi^+ + \pi^+ + \pi^-$$

These are much more complicated processes to deal with theoretically and there is no unambiguous coupling constant like G_V that can be measured.

9.2 THE WEAK INTERACTION

In the foregoing examples of weak decay processes none has been included involving hadrons carrying charm or bottomness. As is clear from Table 7.1 such particles (e.g. the D and B mesons) do undergo weak decays but, for simplicity of presentation, we do not for the moment include them in the discussion.

Consider first the semileptonic decays. These can easily be interpreted in terms of the quark structure of hadrons. Thus neutron decay

$$n(udd) \rightarrow p(uud) + e^- + \tilde{v}_e$$

simply involves the transformation of a d quark into a u quark, i.e.

$$d \rightarrow u + e^- + \tilde{v}_e \qquad (9.6a)$$

Further, this process, in the form

$$d + \tilde{u} \rightarrow e^- + \tilde{v}_e \qquad (9.6b)$$

(note that creation of a particle is equivalent to annihilation of the antiparticle) can similarly account for the decay of π^-:

$$\pi^-(d\tilde{u}) \rightarrow e^- + \tilde{v}_e$$

Including a quark weak interaction with the $\mu^- \tilde{v}_\mu$ pair as well as the $e^- \tilde{v}_e$ pair can account for the $\Delta S = 0$ semileptonic processes involving muons.

Turning now to the $\Delta S = \pm 1$ semileptonic processes, the decay

$$\Lambda^0(sud) \rightarrow p(uud) + e^- + \tilde{v}_e$$

for example, can be accounted for interms of the quark decay process

$$s \rightarrow u + e^- + \tilde{v}_e \qquad (9.7)$$

and so on.

Clearly quarks and leptons must feature in pairs so as to conserve baryon and lepton number. To symbolize the foregoing ideas we therefore introduce a 'current' operator $J(ab)$ which has the property of creating and annihilating particles (or antiparticles) a and b so as to conserve baryon and lepton number. More specifically it enables the following processes:

$$vacuum \rightarrow a + \tilde{b}$$
$$\tilde{a} \rightarrow \tilde{b}$$
$$b \rightarrow a$$
$$\tilde{a} + b \rightarrow vacuum$$

In terms of such currents which, in a full theory, are combinations of 4-component relativistic polar (V) and axial (A) vectors, the interaction responsible for μ^--decay can now be written

$$H_\mu = \frac{G_\mu}{\sqrt{2}} J(v_\mu \mu^-) J(e^- v_e) \qquad (9.8)$$

where G_μ is given in eq. (9.1) and the $\sqrt{2}$ is included for historical reasons. The interaction for μ^+-decay is obtained simply by reversing the ordering of each pair of particles. Using the creation and annihilation properties of the currents it is simple to check that these two interactions do indeed produce the required processes. The current J must contain both polar vector (V) and axial vector (A) parts of equal strength as is indicated by the equality in magnitude of $G_{\mu V}$ and $G_{\mu A}$ (section 9.1). It is this equality which leads to the emission of neutrinos with helicity $\lambda = -1$.

J is referred to as a current since it has similar properties to the electromagnetic current operator J_{em}. The latter, however, only has a polar vector component and, in addition, does not have the property of changing the charges of the particles involved. It simply enables a charged particle to change from one energy–momentum state to another with the emission or absorption of a photon. For this reason, J_{em} is called a **neutral** current whilst the weak current, which can bring about a change of charge, is called a **charged** current.

The semileptonic quark interactions responsible for $\Delta S = 0$ and $\Delta S = \pm 1$ decays (eqs (9.6) and (9.7)) can be written in a similar fashion:

$$H_{\Delta S=0} = \frac{G_V(\Delta S = 0)}{\sqrt{2}}[J(ud)J(e^-v_e) + J(du)J(v_e e^-)] \qquad (9.9)$$

$$H_{\Delta S=\pm 1} = \frac{G_V(\Delta S = \pm 1)}{\sqrt{2}}[J(us)J(e^-v_e) + J(su)J(v_e e^-)] \quad (9.10)$$

where the first term in each pair of square brackets is responsible for $e^-\tilde{v}_e$ emission and the second for e^+v_e emission. The foregoing formulation assumes that $G_A = -G_V$ for the above quark currents just as for the lepton current and we shall see later that this nevertheless leads to an inequality, as observed, when considering the decay of the neutron itself.

The difference in the values of the coupling constants for the three interactions (eqs (9.8), (9.9) and (9.10)) led Cabbibo in 1963 to suggest that the overall weak interaction strength G_F (referred to as the Fermi constant) should be taken equal to G_μ and that it should be 'shared' between the $\Delta S = 0$ and $\Delta S = \pm 1$ quark currents according to the simple prescription

$$G_V(\Delta S = 0) = G_F \cos \theta_C$$
$$G_V(\Delta S = \pm 1) = G_F \sin \theta_C$$

where θ_C is known as the Cabbibo angle and has a value $\simeq 0.22$ in order to account for the values of the coupling constants given in eqs (9.3) and (9.5).

With this formulation it is now possible to write a more general form for the weak interaction including all the processes so far discussed in this section, namely

$$H_{\mathrm{W}} = \frac{G_{\mathrm{F}}}{\sqrt{2}} J^{\dagger} J \tag{9.11}$$

where

$$J = J(\mathrm{ev_e}) + J(\mu\nu_\mu) + J(\mathrm{du})\cos\theta_{\mathrm{C}} + J(\mathrm{su})\sin\theta_{\mathrm{C}}$$

$J^{\dagger}$ is known as the **conjugate** current and is the same as J but with the components of the lepton and quark pairs interchanged. The currents are linear in the particle states so, more elegantly, we can write

$$J = J(\mathrm{ev_e}) + J(\mu\nu_\mu) + J(\mathrm{d' u})$$

where d' ($=\mathrm{d}\cos\theta_{\mathrm{C}} + \mathrm{s}\sin\theta_{\mathrm{C}}$) represents a 'mixed' quark.

This general form for the weak interaction includes all the terms previously discussed but also some additional interactions. For example, the terms involving the product of the two quark currents $J^{\dagger}(\mathrm{du})$ ($= J(\mathrm{ud})$) and $J(\mathrm{su})$ are able to account in principle, and do so in practice, for strangeness-changing non-leptonic weak decay processes. Thus, the product $J^{\dagger}(\mathrm{du})J(\mathrm{su})$ ($= J(\mathrm{ud})J(\mathrm{su})$) which enables the quark process

$$\tilde{\mathrm{s}} \rightarrow \tilde{\mathrm{u}} + \tilde{\mathrm{d}} + \mathrm{u}$$

is easily seen to account for the quark changes in the non-leptonic K^0 decay

$$\mathrm{K}^0(\tilde{\mathrm{s}}\mathrm{d}) \rightarrow \pi^+(\mathrm{u}\tilde{\mathrm{d}}) + \pi^-(\mathrm{d}\tilde{\mathrm{u}})$$

Further, the product proportional to $J^{\dagger}(\mathrm{du})J(\mathrm{du})$ corresponds to a **strangeness-conserving** weak interaction. This, for example, leads to a weak internucleon potential. In magnitude it would be quite undetectable compared with the strong internucleon potential ($V_{\mathrm{weak}} \sim 10^{-7} V_{\mathrm{strong}}$). However, it has the signature that it does not conserve parity and it is gratifying that small parity-violating effects of the right order of magnitude have been observed in nuclear processes such as α- and γ-decay.

This formulation of the weak interaction can readily be extended to include the remaining quarks and to account in general terms for the weak decays of hadrons involving c, b and t quarks. For example, a current involving the charmed quark, c, of the form $J(\mathrm{s' c})$ can be included. Here $\mathrm{s'} = \mathrm{s}\cos\theta_{\mathrm{C}} - \mathrm{d}\sin\theta_{\mathrm{C}}$ is a mixed quark orthogonal to the d′ quark. This then enables this extended form of the weak interaction to account for the decays of the D^+ and D^0 mesons. The complete theory also includes a 'mixed' b′ quark and further small admixtures of b quarks into the d′ and s′ quarks.

Finally, we return briefly to the value of G_A for neutron decay. As things stand using H_W (eq. (9.11)), the coupling constants for d quark decay (eq. (9.6)) are $G_V = -G_A = G_F \cos \theta_C$. The polar vector (V) contribution to neutron decay comes from the 'time' component of the polar vector current included in $J(du)$. This is completely analogous to electric charge which, it will be recalled, is the time component of the electromagnetic 4-vector current. So, just as the charge of a nucleon is simply the sum of the charges of the component quarks (conserved vector current), so G_V for the neutron is simply the value of G_V for the decaying quark, namely $G_F \cos \theta_C$ as is found experimentally. This value for G_V is independent of the details of the quark structure of the nucleon. Like the electromagnetic current the weak polar vector current is said to be conserved.

For G_A, however, the situation is different. Here, it is the three 'spatial' components of the axial vector (A) current included in J that are responsible. These are proportional to the quark spin operator and, in turn, the effective value of G_A for a nucleon depends on how the quark spins are coupled. A simplistic calculation using the quark structure for a nucleon discussed in section 8.2.2 in connection with nucleon magnetic moments leads to the result $G_A/G_V = -5/3 \simeq -1.67$ for neutron decay. This is to be compared with the experimental result -1.27. The disagreement arises because, as has been emphasized in section 8.4, the quarks in a nucleon are highly relativistic. Taking this into account agreement between theory and experiment can be obtained.

To summarize, the weak interaction has been formulated in a simple and elegant fashion (eq. (9.11)) as the interaction between currents governed by a single coupling constant G_F. Further, the different contributions to the currents from leptons and quarks are all of essentially the same form in terms of the d′, s′ and b′ quarks. The next task is to consider the mechanism by which the weak currents interact with each other.

9.3 THE ELECTROWEAK INTERACTION

As mentioned in the last section the electromagnetic current interacts directly with a photon, and the interaction of two electromagnetic currents is effected through the exchange of a virtual photon (Fig. 3.6).

Similarly we have seen (section 8.4) that the quark–quark interaction is propagated by the exchange of gluons (Fig. 8.4). On the contrary, the weak interaction currents have so far been taken to interact at a point (as, for example, in the weak decay processes illustrated in Fig. 6.6). But as early as the 1950s it was conjectured that the interaction between the weak currents might be propagated by the exchange of some form of heavy boson (denoted by W). This is illustrated for some typical weak interaction processes in Fig. 9.1 where it can be seen that in order to conserve charge both positively charged (W^+) and negatively charged (W^-) mesons are needed.

The assumption in the foregoing is that the basic weak interaction coupling is not of the form given in eq. (9.11) but is simply

$$H_W = g(JW + J^\dagger W^\dagger) \tag{9.12}$$

where W and $W^\dagger$ are operators for the creation and annihilation of

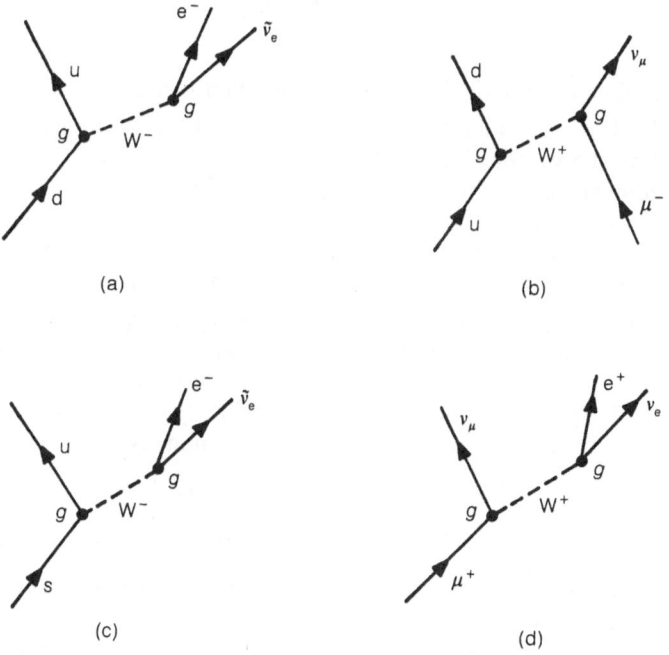

Fig. 9.1 Weak interaction propagated by charged intermediate vector boson ($W^\pm$) for the processes (a) $d \rightarrow u + e^- + \tilde{v}_e$ (as in β^--decay), (b) $u + \mu^- \rightarrow d + v_\mu$ (as in μ^--capture), (c) $s \rightarrow u + e^- + \tilde{v}_e$ (as in Λ^0 semileptonic decay) and (d) $\mu^+ \rightarrow e^+ + v_\mu + v_e$ (μ^+-decay).

charged W bosons and g is now the fundamental weak interaction coupling constant in due course to be related to G_F. Since J is a 4-vector quantity, the $W^{\pm}$ must also be vector bosons (with spin $J = 1$) so that H_W has the required scalar–pseudoscalar properties. The $W^{\pm}$ are frequently referred to as **intermediate vector bosons**.

If this description of the weak interaction is correct, it now has a finite range rather than being a 'contact' interaction as assumed so far. The range is related to the mass M_W of the exchanged boson and will be of the order of magnitude $\hbar/M_W c$ (see the discussion in section 3.4). Since all experimental data are consistent with the weak interaction's being essentially contact in nature, it must therefore be presumed that M_W is very large. This will be seen in due course to be the case.

Turning now to the relation between G_F and g, it is clear from Fig. 9.1 and from the form of H_W (eq. (9.12)) that G_F must be proportional to g^2. Since M_W must also be involved it is easy to show on dimensional grounds that

$$G_F = C \frac{g^2}{M_W{}^2} \left(\frac{\hbar}{c} \right)^2 \tag{9.13}$$

where C is a numerical constant. A detailed calculation gives $C = 1/4\sqrt{2}$. Since the value of G_F is known, eq. (9.13) gives a relationship between the values of g and M_W.

The next step in the development of this theory was taken in the 1960s by Glashow, Weinberg and Salam who proposed that the electromagnetic and weak interactions were both manifestations of a unified electroweak interaction whose overall strength at low energies is determined by the usual electromagnetic coupling constant e. The theory developed, like QED and QCD, is a gauge theory and involves four basic vector fields (or particles), initially massless, denoted by W^+, W^-, W^0 and B. The last two are electrically neutral and are only observed in physical processes as mixtures, namely

$$Z^0 = W^0 \cos \theta_W - B \sin \theta_W$$

$$A = W^0 \sin \theta_W + B \cos \theta_W \tag{9.14}$$

where θ_W is known as the weak mixing angle and has to be determined from experiment.

In the above expressions A represents the usual massless photon whilst the Z^0 is a heavy neutral boson whose mass, M_Z, is related to that of the $W^{\pm}$ by

$$M_Z = M_W / \cos \theta_W \tag{9.15}$$

According to gauge theories all the intermediate bosons must be massless. This appears to be true for the photon and the gluons and the fact that the $W^\pm$ and Z are massive is therefore difficult to understand. However, it turns out that the massless requirement can be evaded by an intriguing proposal known as the **Higgs mechanism**. This brings in its train in prediction that another massive particle, the Higgs boson (with $J = 0$), should exist. So far (1990) it has not been detected and the search for it continues.

Electroweak theory predicts that the weak interaction coupling constant g (eq. (9.12)) for the $W^\pm$ bosons is given by

$$g = \frac{e}{\sqrt{\varepsilon_0} \sin \theta_W} \tag{9.16}$$

so that, using eq. (9.13),

$$G_F = \frac{e^2}{4\sqrt{2}\varepsilon_0 \sin^2 \theta_W M_W^2} \left(\frac{\hbar}{c}\right)^2$$

or, after some rearrangement,

$$\frac{G_F}{(\hbar c)^3} = \alpha \frac{\pi}{\sqrt{2} \sin^2 \theta_W} \frac{1}{(M_W c^2)^2} \tag{9.17}$$

where α is the usual electromagnetic fine structure constant. In the next section the experimental value of $\sin^2 \theta_W$ will be determined and hence, knowing G_F, it becomes possible to obtain a value for M_W from eq. (9.17).

Finally, as we have seen in eq. (9.14), electroweak theory predicts the existence of an additional boson – the Z^0. This is a massive, neutral particle and is coupled to a weak neutral current with

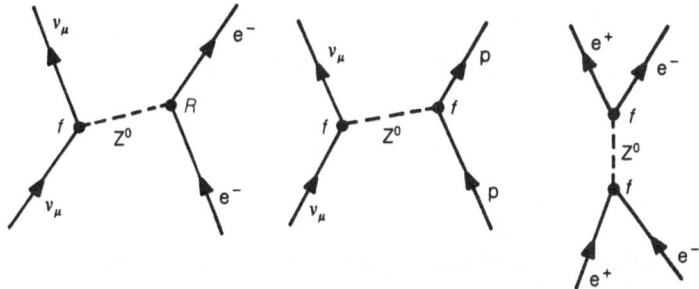

Fig. 9.2 Examples of weak neutral current interaction processes.

coupling constant $f = e/\sqrt{\varepsilon_0} \cos \theta_W \sin \theta_W$. This coupling, in turn, allows additional weak interaction processes such as

$$\nu_\mu + e^- \rightarrow \nu_\mu + e^-$$
$$\nu_\mu + p \rightarrow \nu_\mu + p$$
$$e^+ + e^- \rightarrow e^+ + e^- \qquad (9.18)$$

These are illustrated in Fig. 9.2. In the next section we shall see that such processes have been observed and, indeed, lead to experimental values for θ_W. Thus, the next step is to see to what extent experiment does agree with the predictions of electroweak theory.

9.3.1 Experimental Confirmation of the Electroweak Interaction

The unified electroweak theory just described accounts excellently for the features of the weak interaction encountered so far. But, in addition, it requires the existence of a neutral boson – the Z^0 – and associated neutral current processes such as those given in eq. (9.18). Experimental evidence for these processes first emerged in 1973 and this will now be briefly described.

The original experiments used protons from the CERN proton synchrotron bombarding a metal target to produce π^+ and π^- mesons. By use of appropriate magnetic fields, collimated beams of π^+ or π^- mesons were produced. These mesons then decayed (Table 7.1) after passing though a drift region into $\mu^+ + \nu_\mu$ and $\mu^- + \tilde{\nu}_\mu$ respectively. At the end of the drift region was a thick wall through which the neutrinos easily passed whilst the charged muons and other residual particles were stopped and absorbed. Thus the outcome was a ν_μ or $\tilde{\nu}_\mu$ beam with energy around 1 GeV. A search was then made using a large heavy liquid bubble chamber (called Gargamelle) for evidence of neutrino (or antineutrino) scattering processes of the form

$$\nu_\mu + \text{nucleus} \rightarrow \nu_\mu + \text{hadrons}$$

or

$$\nu_\mu + e^- \rightarrow \nu_\mu + e^-$$

and similar processes involving antineutrinos. Processes such as these were indeed detected with cross-sections of the order 10^{-18} b, a very small value compared with typical strong interaction cross-sections (~ 1 mb), indicating the weakness of the weak interaction.

The theoretical value for the cross-section depends on the value of

θ_W and, hence, by comparing theory and experiment a value for θ_W could be determined. A recent (1988) value taking into account not only neutrino scattering experiments but also measurements of processes such as the third in eq. (9.18) is

$$\sin^2 \theta_W = 0.230 \pm 0.0048 \qquad (9.19)$$

With this value, use of eq. (9.17) and the value of G_F now enables M_W and M_Z to be determined. Using

$$G_F = G_\mu = 1.166 \times 10^{-5} (\hbar c)^3 \, \text{GeV}^{-2} \quad \text{(see eq. (9.1))}$$
$$\alpha = 7.297 \times 10^{-3}$$

gives

$$M_W = \frac{37.29}{\sin \theta_W} \, \text{GeV}/c^2 = 77.75 \, \text{GeV}/c^2$$

In turn, use of eq. (9.15) gives

$$M_Z = \frac{M_W}{\cos \theta_W} \, \text{GeV}/c^2 = 88.60 \, \text{GeV}/c^2$$

Taking into account some small corrections, and allowing for experimental and theoretical uncertainties, a recent (1988) assessment gives

$$M_W = 80.2 \pm 1.1 \, \text{Gev}$$
$$M_Z = 91.6 \pm 0.9 \, \text{GeV} \qquad (9.20)$$

These values are very large ($\sim 100 \times$ proton mass) and imply that the non-locality of the weak interaction is of the order $\hbar/M_W c \sim 10^{-18}$ m, i.e. it is effectively a contact interaction as observed experimentally. Since $G_F \propto M_W^{-2}$ (eq. (9.13)), the large mass values also explain why the interaction is weak even though the basic coupling strength is the same as for the electromagnetic interaction.

Clearly electroweak theory would become firmly established if the W and Z bosons could be detected experimentally. This was achieved in 1983 by Rubbia *et al.* using the CERN Super Proton Synchrotron (SPS) which accelerates protons to energies around 270 GeV. However, as was shown in section 5.1.1 for particles striking a fixed target, the energy in the centre-of-mass system is significantly less than the energy of the bombarding particles. The calculation there was non-relativistic whilst for SPS energies a relativistic formulation is needed (for example Turner (1984)). For protons of energy E colliding with target protons, the centre-of-mass energy E_{cm} released is given approximately by

$$E_{cm} \simeq (2Em_p c^2)^{1/2}$$

and, for $E = 270 \, GeV$, this gives $E_{cm} \simeq 23 \, GeV$. This is far too small to lead to the production of W and Z bosons. However, a method was devised in which antiprotons were accelerated in the SPS in the opposite direction to the protons and the two beams then brought into collision. For such a colliding beam experiment the centre of mass is fixed in the laboratory frame of reference, and the energy released is simply $2E = 540 \, GeV$. Thus there is plenty of energy for the creation of W and Z bosons.

The creation takes place through weak interaction processes involving the component quarks in the proton (p) and antiproton (p̃) such as

$$\tilde{u} \, (from \, \tilde{p}) + d \, (from \, p) \rightarrow W^-$$
$$\tilde{u} \, (from \, \tilde{p}) + u \, (from \, p) \rightarrow Z^0$$

The W^- and Z^0 are predicted to decay in time $\sim 10^{-23} \, s$ into either quarks or leptons, for example

$$W^- \rightarrow \tilde{u} + d$$
$$\rightarrow e^- + \tilde{\nu}_e$$
$$\rightarrow \mu^- + \tilde{\nu}_\mu$$
$$Z^0 \rightarrow \tilde{u} + u$$
$$\rightarrow e^+ + e^-$$
$$\rightarrow \mu^+ + \mu^-$$

The decay quarks immediately produce jets which are indistinguishable from jets produced in strong interaction processes (section 8.4) resulting from the pp̃ collisions. However, there is no confusion over the two leptons which, assuming that the W^- and Z^0 are produced at rest in the laboratory frame, will be produced 'back to back' each with energy around $M_W c^2/2$ or $M_Z c^2/2$, i.e. 40–45 GeV. Extremely complicated and large (15–20 m) detectors were set up to detect leptons in this energy range and the creation of W and Z bosons was clearly established. Careful measurements of the lepton energies enabled the boson masses to be measured and recent (1990) averaged experimental results give

$$M_W = 80.49 \pm 0.67 \, GeV/c^2 \quad M_Z = 91.49 \pm 1.39 \, GeV/c^2$$

in very good agreement with the theoretical predictions given in eq. (9.20).

The electroweak theory is clearly very successful and it is not surprising that its originators (Glashow, Weinberg and Salam) together with Rubbia and Van der Meer, who were the principal experimental investigators of the theory, were all awarded Nobel prizes.

Finally, a comment about the strangeness-changing neutral current. The neutral current involves a term $J(d'd')$ where the d' is the mixed quark ($= d \cos \theta_C + s \cos \theta_C$) defined in section 9.2. The current would then contain cross-terms such as $J(ds) \sin \theta_C \cos \theta_C$ which could lead to strangeness-changing neutral processes, for example, $\Delta^0 \rightarrow n + \nu_e + \tilde{\nu}_e$, which are known not to occur. However, there is another contribution from $J(s's')$ where $s' = s \cos \theta_C - d \sin \theta_C$ which contains exactly the same cross-term but with the opposite sign. The two therefore cancel and the difficulty is removed. It will be recalled that the s' was introduced to couple to the charmed quark and originally this quark was hypothesized in order to achieve cancellations of the type just discussed. It was because of its beneficial effect in this connection that it was referred to as charmed!

9.4 CP VIOLATION

In section 7.3.3 it was pointed out that all interactions appeared to be invariant under the combined operations of charge conjugation (C) and reflection of axes (the parity operation P). This is generally referred to as CP invariance. There is, in fact, a more fundamental invariance under the three operations of charge conjugation, parity and time reversal (T), known as CPT invariance, which is believed to hold absolutely. The so-called CPT theorem holds if the very general assumptions underlying relativistic quantum mechanics are correct. There is no evidence that these assumptions are invalid and if they turned out to be so this would completely undermine our basic thinking.

By time reversal is meant reflection (i.e. changing the sign) of the time coordinate just as the parity operation involves changing the sign of the three space coordinates. If CP invariance holds, then it follows from the CPT theorem that interactions are also invariant under time reversal. Simplistically, this means that a film of any elementary particle process run backwards should still show a physically allowable process.

The weak interaction and unification

There is so far no direct experimental evidence of non-invariance under time reversal. For example, a comparison of the forward and backward rates of the nuclear reaction

$$p + {}^{27}Al \rightleftharpoons \alpha + {}^{24}Mg$$

has shown that T violation is less than $\simeq 5 \times 10^{-4}$. Limits have also been put on such violation from studies of angular correlations in nuclear β-decay.

However, in 1964 an experiment on K^0 decay showed unequivocal violation of CP invariance and thus, implicitly through the CPT theorem, that T violation was taking place. The K^0 and its antiparticle, the $\tilde{K}^0$, are special in the sense that they 'mix' with one another via the weak interaction. The K^0 has $S = +1$ and the $\tilde{K}^0$ has $S = -1$. Each can transform via the weak interaction into, for example, a π^+ and a π^- (section 9.2). This means that, as stated above, the K^0 can transform into the $\tilde{K}^0$ and vice versa, as shown in Fig. 9.3, so that physically mixed states will be expected to be observed.

The mixed states of importance for our considerations are eigenstates of the CP operator and are defined as follows

$$K_S^0 = \frac{1}{\sqrt{2}}\left(K^0 + \tilde{K}^0\right) \qquad K_L^0 = \frac{1}{\sqrt{2}}\left(K^0 - \tilde{K}^0\right) \qquad (9.21)$$

By convention there is a phase factor equal to -1 when the charge conjugation operator C operates on the K^0 and $\tilde{K}^0$, i.e.

$$CK^0 = -\tilde{K}^0 \quad \text{and} \quad C\tilde{K}^0 = -K^0 \qquad (9.22)$$

Thus, since the K^0 and $\tilde{K}^0$ have odd parity, i.e.

$$PK^0 = -K^0 \quad \text{and} \quad P\tilde{K}^0 = -\tilde{K}^0 \qquad (9.23)$$

it follows that

$$CPK_S^0 = K_S^0 \quad \text{and} \quad CPK_L^0 = -K_L^0 \qquad (9.24)$$

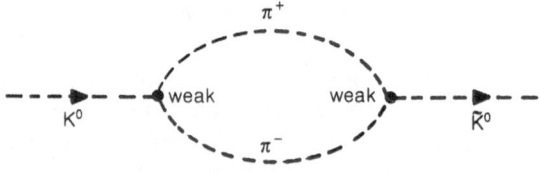

Fig. 9.3 Transformation of K^0 into $\tilde{K}^0$ via the weak interaction.

If CP invariance holds, the K_S^0, having CP $= +1$, can only decay into two pions ($\pi^+\pi^-$ or $\pi^0\pi^0$) which also have CP $= +1$. That CP $= +1$ follows most simply for two π^0s. In this case the relative orbital angular momentum, l, of the two π^0s must be even so that the wave function is symmetrical under exchange of the two (identical) bosons. Thus since

$$C\pi^0 = \pi^0 \quad \text{(the } \pi^0 \text{ is its own antiparticle)}$$

and

$$P\pi^0 = -\pi^0 \quad \text{(since pions have odd parity)} \qquad (9.25)$$

it follows that $CP(\pi^0\pi^0) = (-1)^2(-1)^l(\pi^0\pi^0) = \pi^0\pi^0$ and CP $= 1$. A slightly more complicated argument gives the same result for the $\pi^+\pi^-$ combination.

On the contrary, it can be shown that three pions have CP $= -1$ and so, for CP invariance, the K_L^0 must decay to three pions as (usually) observed experimentally (Table 7.1). Because the decay of a K_S^0 into two pions releases much more kinetic energy than in a three-pion decay its lifetime is significantly shorter than that of the K_L^0. In Table 7.1 it can be seen that $\tau(K_S^0) = 0.89 \times 10^{-10}$ s whilst $\tau(K_L^0) = 5.18 \times 10^{-8}$ s.

Using this difference in lifetimes, Cronin and Fitch in 1964 were able to make careful measurements on the decay of K_L^0 mesons. They used the process

$$\pi^- + p \rightarrow \Lambda^0 + K^0$$

to produce a K^0 beam. The K^0 are, of course, not eigenstates of CP but, using eqs (9.21), the K^0 can be written as a superposition of K_S^0 and K_L^0 as follows

$$K^0 = \frac{1}{\sqrt{2}}\left(K_S^0 + K_L^0\right) \qquad (9.26)$$

The K^0 beam must therefore be regarded as made up initially of 50% K_S^0 and 50% K_L^0. As has been seen, the K_S^0 mesons decay some 500 times faster than the K_L^0 and so, allowing the K^0 beam to drift freely for a sizeable distance, as was done by Cronin and Fitch, an end beam consisting only of K_L^0 mesons remains, the K_S^0 mesons having all decayed. If CP invariance holds the K_L^0 should decay into three pions. It was found, however, that around 0.2% of the decays were into two pions, indicating a small violation of CP invariance. This can be interpreted as due to a small admixture of K_S^0 into the K_L^0

by a CP-violating interaction with an amplitude $\eta \simeq 2 \times 10^{-3}$. This result has been confirmed by subsequent experiments and CP violation has also been observed in leptonic K^0 decay but, to date (1990), it has not been observed in any other process.

The origin of this violation is not clear, although a number of theories to account for it have been suggested. One proposal is that it results from imaginary parts in the mixing amplitudes of the quark components in the s' and b' superpositions discussed in section 9.2. Another is that there is a very weak (superweak) $\Delta S = 2$ CP-violating interaction which can couple the K^0 and $\tilde{K}^0$ together. If the latter is correct, because of its great weakness, its effects could only be observed within the $K^0\tilde{K}^0$ system.

One important area of experimental research aimed at clarifying the situation is to look for an electric dipole moment ($d_n \sim er$ and measured in units of e m) of the neutron. Like the neutron magnetic moment this will be proportional to the neutron spin operator J. However, under the parity operation d_n changes sign whilst J does not and under the time reversal operation J changes sign whilst d_n does not. Thus, for the proportionality to hold both P and T violation must occur and so a finite value for the dipole moment would, in particular, indicate T violation. Attempts to measure d_n using neutrons confined by electromagnetic fields ('bottled' neutrons) have so far given $d_n < 1 \times 10^{-27}e$ m. Theoretically the order of magnitude expected for d_n is

$$d_n \sim (e \times 10^{-15})(10^{-5})\varepsilon \, \text{m}$$
$$\sim \varepsilon 10^{-20} e \, \text{m} \tag{9.27}$$

where the quantity in the first pair of parentheses is the basic size of a nucleon dipole moment ($\simeq$ nucleon charge (e) × nucleon size (10^{-15}m)). The quantity in the second pair of parentheses is the relative strength of the weak interaction compared with the strong interaction (e.g. section 6.5.2) and is included to allow for the necessary P violation. The third term, ε, represents the strength of the T-violating interaction. Comparing the experimental limit with eq. (9.27) gives $\varepsilon < 10^{-7}$. This value is much less than the CP (i.e. T) violation ($\eta \simeq 2 \times 10^{-3}$) observed in K_L^0 decay but is consistent with the two theories mentioned above. It does, however, rule out some other proposals that have been made. The search for an experimental value for d_n continues.

9.5 GRAND UNIFIED THEORIES (GUTS)

We conclude by briefly mentioning further developments in elementary particle physics that are in train. They are largely determined by the desire to develop some all-embracing theory of the various aspects of elementary particle physics that have been discussed in the last three chapters.

The extremely successful unification of the weak and electromagnetic interactions discussed in section 9.3 has stimulated the search for an enlargement of this unification to include also the strong interaction. Physicists have been encouraged by the fact that electroweak theory is a gauge theory based on the $SU(2) \otimes U(1)$ symmetry groups and that quantum chromodynamics – the basic theory of the strong interaction – is also a gauge theory based on the $SU(3)$ symmetry group. Further, there is a clear correspondence between the three generations of leptons $(ev_e), (\mu v_\mu), (\tau v_\tau)$ and the three generations of quarks (du), (sc), (tb).

Of course, at the energies at which experimental studies of these interactions have been made so far we have seen that the basic electroweak symmetry is broken, resulting in three massive gauge bosons $(W^\pm, Z^0)$, on the one hand, and a massless boson (the photon, γ) on the other. In addition, the strengths of the different interactions as measured, for example, by α and α_s (section 8.4) are very different. However, as noted in section 8.4, as the separation between quarks decreases below 10^{-15} m there is a corresponding decrease in α_s (asymptotic freedom). Remembering that the de Broglie wavelength for high energy particles is given by $\lambda = h/p \simeq hc/E$, where E is the particle energy, it follows that a decrease in α_s should become apparent when E increases well above 1 GeV (for which $\lambda \simeq 10^{-15}$ m). Detailed theoretical investigation of the dependence of α_s on energy, and also of the electroweak coupling constants, shows that they all become equal to one another when $E \simeq 10^{15}$ GeV, well above the Z, W and quark masses, corresponding to a separation $\simeq 10^{-30}$ m. This suggests that at such an energy the strong, electromagnetic and weak interactions become essentially the same and that the differences observed in low (relatively speaking) energy laboratory experiments result from spontaneous breaking of the overall symmetry at these energies.

No attempt will be made to describe the different unified theories that have been developed. Suffice it to say that they involve larger symmetry groups which, in turn, lead to the introduction of new particles. For example gauge bosons called leptoquarks have been postulated which, among other things, allow the decays

$$p \rightarrow \pi^+ + \tilde{v}_e$$
$$\rightarrow e^+ + \pi^0$$

which violate both lepton and baryon number conservation. The predicted lifetime, τ, for the proton is at least 10^{30} years, depending on the details of the theory. Experiments underway to detect this decay using gigantic underground tanks of water in which, for example, the Cerenkov radiation of a decay positron is measured have so far suggested that $\tau > 10^{32}$ years. A bonus from these theories is that they have been able to account for the third-integral nature of quark charge and for the equality in magnitude of the proton and electron charges.

Another theory, referred to as 'supersymmetry', requires the existence of spin $\frac{1}{2}$ particles corresponding to each of the usual gauge bosons. They are referred to collectively as 'sparticles' and individually as 'photinos', 'gluinos' etc. So far none has been identified in accelerator experiments, implying that, if they exist, they have very large masses.

Finally, it should be remarked that the ultimate target is to unify the elementary particle interactions with the gravitational interaction. Dimensionally, using the key constants for quantum mechanics ($\hbar$), relativity (c) and gravitation (the gravitational constant $G = 6.672 \times 10^{-11} \, \mathrm{N \, m^2 \, kg^{-2}}$), the characteristic mass, known as the Planck mass, for such a unification is given by

$$M_p = \left(\frac{\hbar c}{G}\right)^{1/2} = 2.175 \times 10^{-8} \, \mathrm{kg} = 1.22 \times 10^{19} \, \mathrm{GeV}/c^2$$

This corresponds to four orders of magnitude higher energy than that at which unification of elementary particle interactions takes place. Any theory encompassing gravitation requires a further gauge boson – the graviton – with $J = 2$ and possibly a 'gravitino' with $J = \frac{3}{2}$. Such theories also run into the difficulty that because of quantum fluctuations of range $\hbar/M_p c \simeq 1.7 \times 10^{-35} \, \mathrm{m}$ (the Planck length) satisfactory progress cannot be made using the concept of point particles. The solution to this problem has involved the representation of particles by finite one-dimensional strings sometimes forming closed loops, and such theories appear to have great potential for the future.

References

Arima, A. and Iachello, F. (1975). *Phys. Lett.*, **35**, 1069.

Bohr, A. and Mottelson, B. R. (1969). *Nuclear Structure*, W. A. Benjamin Inc., New York.

Brown, G. E. and Jackson, A. D. (1976). *The Nucleon–Nucleon Interaction*, North-Holland/American Elsevier, Amsterdam.

Burcham, W. E. (1988). *Elements of Nuclear Physics*, Reprint, Longman, London.

Green, A. E. S. (1954). *Phys. Rev.*, **95**, 1006.

Kenyon, I. R. (1987). *Elements of Nuclear Physics*, Routledge and Kegan Paul, London.

Particle Data Group (1988). *Phys. Lett.*, **204B**, 1.

Turner, R. E. (1984). *Relativity Physics*, Routledge and Kegan Paul, London.

Weisskopf, V. F. (1957). *Phys. Today*, **14**(7), 18.

Further reading

The following are a few recently published texts on nuclear and particle physics which extend the coverage of this book.

Burcham, W. E. (1988). *Elements of Nuclear Physics*, Reprint, Longman, London.

Hughes, I. S. (1985). *Elementary Particles*, 2nd edn, Cambridge University Press, Cambridge.

Jelley, N. A. (1990), *Fundamentals of Nuclear Physics*, Cambridge University Press, Cambridge.

Jones, G. A. (1987). *The Properties of Nuclei*, 2nd edn, Oxford University Press, Oxford.

Kane, G. (1987). *Modern Elementary Particle Physics*, Addison-Wesley, Wokingham.

Kenyon, I. R. (1987). *Elementary Particle Physics*, Routledge and Kegan Paul, London.

Satchler, G. R. (1990). *Introduction to Nuclear Reactions*, 2nd edn, MacMillan, London.

Table of physical constants

Constant	Symbol	Value
speed of light	c	$2.998 \times 10^8 \, \text{m s}^{-1}$
permeability of vacuum	μ_0	$4\pi \times 10^{-7} \, \text{H m}^{-1}$
permittivity of vacuum	$\varepsilon_0 = 1/\mu_0 c^2$	$8.854 \times 10^{-12} \, \text{F m}^{-1}$
elementary charge	e	$1.602 \times 10^{-19} \, \text{C}$
gravitational constant	G	$6.673 \times 10^{-11} \, \text{N m}^2 \text{kg}^{-2}$
Atomic mass unit	u	$1.661 \times 10^{-27} \, \text{kg}$
mass		
of electron	m_e	$9.109 \times 10^{-31} \, \text{kg}$
of proton	m_p	$1.673 \times 10^{-27} \, \text{kg}$
of neutron	m_n	$1.675 \times 10^{-27} \, \text{kg}$
energy equivalence of mass		
of electron	$m_e c^2$	$0.511 \, \text{MeV}$
of proton	$m_p c^2$	$938.272 \, \text{MeV}$
of neutron	$m_n c^2$	$939.566 \, \text{MeV}$
Planck constant	h	$6.626 \times 10^{-34} \, \text{J s}$
$h/2\pi$	$\hbar$	$1.055 \times 10^{-34} \, \text{J s}$
fine structure constant	$\alpha = e^2/4\pi\varepsilon_0 \hbar c$	7.297×10^{-3}
Bohr magneton	$\mu_B = e\hbar/2m_e$	$9.274 \times 10^{-24} \, \text{J T}^{-1}$
nuclear magneton	$\mu_N = e\hbar/2m_p$	$5.051 \times 10^{-27} \, \text{J T}^{-1}$
magnetic moment		
of electron	μ_e	$9.285 \times 10^{-24} \, \text{J T}^{-1}$
of proton	μ_p	$1.411 \times 10^{-26} \, \text{J T}^{-1}$
		$= 2.793 \, \mu_N$
of neutron	μ_n	$-0.966 \times 10^{-26} \, \text{J T}^{-1}$
		$= -1.913 \mu_N$
electronvolt	eV	$1.602 \times 10^{-19} \, \text{J}$
	$\text{keV} = 10^3 \, \text{eV}$	$1.602 \times 10^{-16} \, \text{J}$
	$\text{MeV} = 10^6 \, \text{eV}$	$1.602 \times 10^{-13} \, \text{J}$
	$\text{GeV} = 10^9 \, \text{eV}$	$1.602 \times 10^{-10} \, \text{J}$
	$\text{TeV} = 10^{12} \, \text{eV}$	$1.602 \times 10^{-7} \, \text{J}$
barn	b	$10^{-28} \, \text{m}^2$
	$\text{mb} = 10^{-3} \, \text{b}$	$10^{-31} \, \text{m}^2$
	$\mu\text{b} = 10^{-6} \, \text{b}$	$10^{-34} \, \text{m}^2$
	$\text{nb} = 10^{-9} \, \text{b}$	$10^{-37} \, \text{m}^2$

Problems

(Relevant fundamental constants are given on p. 195)

1 The nuclear atom

1.1 What is the distance of closest approach of 5.0 MeV and 7.0 MeV α-particles to nuclei of the following elements Ag($Z = 47$), Au ($Z = 79$) and Pb ($Z = 82$)?

1.2 Use the uncertainty relation to estimate (in MeV) the minimum kinetic energy possible for (i) an electron confined to a sphere of diameter 2 fm and (ii) a nucleon confined similarly. What are the implications of these estimates for deciding whether the nucleus is composed of protons and electrons or protons and neutrons?

1.3 If nuclei were constituted from protons and electrons only (as originally postulated), determine whether the nuclear spin would be integer or half-integer in the following circumstances:

A	Z	Spin
even	even	?
even	odd	?
odd	even	?
odd	odd	?

and compare your results with those obtained when the nucleus is constituted from protons and neutrons.

2 General properties of the nucleus

2.1 The measured radii of nuclei are well represented by the expression $R = 1.1A^{1/3}$ fm where A is the mass number. Find the volume per nucleon in fm^3 and show that it is the same as that of a sphere of radius 1.1 fm.

2.2 The mass of the atom $^{20}_{10}$Ne is 19.9924 u. Find the nuclear binding energy in MeV. ($m_n = 1.0087$ u; $m_H = 1.0078$ u.)

2.3 How much energy is required to remove (i) one neutron or (ii) one proton from $^{16}_{8}$O? (The masses of relevant neutral atoms are $M(^{15}_{7}\text{N}) = 15.0001$ u, $M(^{15}_{8}\text{O}) = 15.0030$ u and $M(^{16}_{8}\text{O}) = 15.9949$ u.)

3 The internucleon potential

3.1 Assuming that the deuteron (binding energy $= B \simeq 2$ MeV) has radius $R \simeq 1$ fm, use the uncertainty relation to estimate the least kinetic energy of a component nucleon and hence deduce an order of magnitude for the internucleon potential energy.

3.2 Low energy neutron–proton scattering data can be accounted for very approximately by representing the internucleon potential by an attractive finite square well of range $a = 2$ fm and depth $V_0 = 35$ MeV in the $^3S_1$ state and $V_0 = 15$ MeV in the $^1S_0$ state.

This potential can be represented by an expression of the form

$$V = A + Bs_1 \cdot s_2 \qquad r \leqslant a$$
$$= 0 \qquad r > a$$

Determine the values of A and B. (Hint: relate the expectation value of $s_1 \cdot s_2$ to that of J^2 where $J = s_1 + s_2$.)

3.3 If the internucleon potential is charge independent, the 'mirror' nuclei $^{35}_{18}$A and $^{35}_{17}$Cl can be considered to have essentially the same nuclear structure. The 5.97 MeV energy difference between their ground states can then be attributed to the difference in their Coulomb energies and between the masses of the neutron and proton. Assuming these nuclei to have a uniform, spherical charge distribution estimate their radius.

3.4 Estimate the range of the parts of the internucleon potential stemming from the exchange of the following mesons: (i) η meson ($m_\eta \simeq 550$ MeV/c^2), (ii) ρ meson ($m_\rho \simeq 770$ MeV/c^2) and (iii) two pions ($m_\pi \simeq 270$ MeV/c^2).

4 Models of nuclear structure

4.1 Assuming that for a nucleus $^A_Z X$ the nuclear charge is spread uniformly throughout the nucleus (assumed spherical and of radius $R = 1.1A^{1/3}$ fm), show that the Coulomb energy is given by

$$E_C = 0.79 Z^2 A^{-1/3} \text{ MeV}$$

(Compare with the value of a_C in eq. (4.1) and below.)

4.2 By taking $\delta(A, Z) = 0$ in the Weizsäcker semi-empirical formula (eq. (4.1)) for the mass $M(A, Z)$ of a nucleus, obtain the expression (eq. (4.4))

$$Z = \frac{A}{1.97 + 0.015 A^{2/3}}$$

for the value of Z which corresponds to the most stable nucleus for a given A.

Compare the predictions of this formula with the Segrè chart of stable nuclei given in Fig. 2.3.

4.3 The spins and parities of the following nuclei are as indicated:

$$^{15}_{7}\text{N}(\tfrac{1}{2}^-) \qquad ^{21}_{10}\text{Ne}(\tfrac{3}{2}^+) \qquad ^{39}_{19}\text{K}(\tfrac{3}{2}^+)$$

$$^{51}_{23}\text{V}(\tfrac{7}{2}^-) \qquad ^{73}_{32}\text{Ge}(\tfrac{9}{2}^+) \qquad ^{207}_{82}\text{Pb}(\tfrac{1}{2}^-)$$

Account for them in terms of the nuclear shell model.

4.4 The nuclei $^{113}_{49}\text{In}$, $^{121}_{51}\text{Sb}$, $^{123}_{51}\text{Sb}$ and $^{123}_{52}\text{Te}$ have spins 9/2, 5/2, 7/2 and 1/2 respectively with magnetic moments 5.5 μ_n, 3.4μ_n, 2.6μ_n and $-0.7\mu_n$. Interpret these results in terms of shell model states for the odd nucleon.

4.5 Prove that for a single-particle spin–orbit potential

$$V_{so} = \frac{\lambda}{\hbar^2} \boldsymbol{L} \cdot \boldsymbol{s}$$

where λ is a constant, the energy splitting of the states $j = l \pm 1/2$ is given by $\lambda (2l + 1)/2$.

In the level diagram for ^{17}O and ^{17}F (Fig. 2.4) the $\tfrac{5}{2}^+$ ground state and $\tfrac{3}{2}^+$ excited state are separated by $\simeq 5$ MeV. Assuming that these are $1d_{5/2}$ and $1d_{3/2}$ single-particle states (section 4.2.1) determine the sign and value of λ for these nuclei.

4.6 Show that the rotational energy levels of an even–even nucleus (mass number A) assumed to be rigid and spherical with radius

$R = 1.1A^{1/3}$ fm can be represented by $BJ(J + 1)$ with $J = 0, 2, 4, \ldots$
where

$$B \simeq 43A^{-5/3} \text{ MeV}$$

4.7 The nucleus ^{176}Hf has ground and excited rotational state energy
levels as follows: $0^+(0.000 \text{ MeV})$, $2^+(0.088 \text{ MeV})$, $4^+(0.290 \text{ MeV})$.
What would you predict for the energy of the 6^+ state?

Deduce the moment of inertia for this nucleus and compare it with
the value for rigid rotation (assuming the nucleus to be spherical).

5 Nuclear reactions

5.1 The binding energy of the deuteron (^{2}H) is 2.22 MeV. Find the
minimum energy in the laboratory system that a proton must have in
order to initiate the reaction

$$p + d \rightarrow p + p + n$$

5.2 In the heavy ion reaction

$$^{48}\text{Ca} + {}^{16}\text{O} \rightarrow {}^{49}\text{Sc} + {}^{15}\text{N}$$

the Q-value is -7.83 MeV. What is the minimum kinetic energy of
bombarding ^{16}O ions to initiate the reaction?

At this energy, estimate the orbital angular momentum (in units of
$\hbar$) of the ions for a 'grazing' collision (take $R = 1.1A^{1/3}$ fm).

5.3 A beam of slow neutrons impinges on a boron foil (density of
boron $= 2.5 \times 10^3 \text{ kg m}^{-3}$) and 95% are absorbed. If the reaction
cross-section is $\sigma = 4000$ b, calculate the thickness of the foil.

5.4 Complete the following nuclear reactions indicating the appropriate compound nucleus:

$$^{35}_{17}\text{Cl} + ? \rightarrow ? \rightarrow {}^{32}_{16}\text{S} + {}^4_2\text{He}$$

$$^{10}_5\text{B} + ? \rightarrow ? \rightarrow {}^7_3\text{Li} + {}^4_2\text{He}$$

$$^{23}_{11}\text{Na} + \text{p} \rightarrow ? \rightarrow {}^{20}_{10}\text{Ne} + ?$$

$$^{23}_{11}\text{Na} + {}^2_1\text{H} \rightarrow ? \rightarrow {}^{24}_{11}\text{Na} + ?$$

5.5 A nuclear reactor for producing neutron beams is refuelled with
10 kg of enriched uranium which contains initially 91% of the fissile
isotope ^{235}U. The reactor operates continuously at a constant
thermal power of 57 MW until the quantity of ^{235}U is reduced to 70%
of what was there initially, at which point the ^{235}U is too dilute for the
chain reaction to continue. At what rate are neutrons released and

what is the longest possible period of continuous running? (Assume that the fission of one ^{235}U nucleus releases on average 2.4 neutrons and 185 MeV of heat energy.)

5.6 Check that the total energy released (24.7 MeV) in the hydrogen and carbon cycles is the appropriate sum of the energies released in the individual reactions. (Remember that a reaction may need to be invoked more than once.)

6 Alpha, beta and gamma decay

6.1 A thermal neutron beam of intensity 10^{10} neutrons s^{-1} is directed for two days into a cavity in a block of cobalt initially composed of the stable isotope ^{59}Co. The neutrons are all captured, producing ^{60}Co nuclei which are radioactive with a half-life of 5.3 years. What is the activity of the ^{60}Co (γ-ray) source produced?

6.2 The half-life of radon (^{222}Rn) is 3.823 days. What fraction of a freshly separated sample decays in 1 day and 10 days?

6.3 Show that a full evaluation of the integral in eq. (6.16) leads to the result

$$\gamma = \frac{4Ze^2}{4\pi\varepsilon_0 hv}\left\{\cos^{-1}\left(\frac{R}{r_T}\right)^{1/2} - \left[\frac{R}{r_T}\left(1 - \frac{r}{r_T}\right)\right]^{1/2}\right\}$$

Hence show that, with the approximation $R \to 0$, γ is now given by

$$\gamma = \frac{2\pi Ze^2}{4\pi\varepsilon_0 hv}$$

The difference between this and the expression for γ in eq. (6.19) then accounts for the replacement of 16 by 4π referred to in the text following eq. (6.22).

6.4 Calculate the kinetic energy (in MeV) of α-particles emitted in the decay process

$$^{232}U \to {}^{228}_{92}Th + \alpha$$

Remember nuclear recoil, and take $M(^{232}U) = 232.1095$ u, $M(^{228}Th) = 228.0998$ u and $M(\alpha) = 4.0039$ u.

6.5 Using dimensional arguments, prove eq. (6.26).

6.6 In a nucleus the ground and first three excited states have respectively the following values of J^{π}: $\frac{1}{2}^+$, $\frac{11}{2}^-$, $\frac{5}{2}^+$ and $\frac{7}{2}^+$. Draw a notional energy level diagram and identify the multipolarities of the following γ-transitions:

$$\tfrac{7}{2}^+ \to \tfrac{5}{2}^+, \tfrac{7}{2}^+ \to \tfrac{11}{2}^-, \tfrac{5}{2}^+ \to \tfrac{1}{2}^+, \tfrac{11}{2}^- \to \tfrac{1}{2}^+$$

6.7 The first excited state (2^+) of ^{60}Ni decays to the ground state (0^+) emitting a photon. The energy separation of the two states is 1.33 MeV.

(a) What is the multipolarity of the emitted photon?
(b) Allowing for nuclear recoil, by how much does the measured energy of the photon in the laboratory frame of reference differ from 1.33 MeV?
(c) If the lifetime of the decaying state is $\tau \simeq 10^{-14}$ s, estimate the energy width of the state and comment on the possibility of resonant absorption of the emitted photon by another Ni nucleus.
(d) How would the situation be changed if the Ni crystal as a whole recoiled (Mössbauer effect)?

6.8 Prove the expressions given in eqs (6.39) and (6.40) for the Q values in β^--decay and electron capture.

6.9 Assuming that the neutrino has finite mass m_ν show that the β-decay energy spectrum has the following form

$$P(\varepsilon)\,d\varepsilon \propto (\varepsilon^2 - 1)^{1/2}\varepsilon[(\varepsilon_0 - \varepsilon)^2 - m_\nu^2/m_e^2]^{1/2}(\varepsilon_0 - \varepsilon)\,d\varepsilon$$

(compare with eq. (6.46) *et seq.*).

By writing $\varepsilon = \varepsilon_m - x$ where ε_m is the maximum energy available to the electron, show that for $x \to 0$ the energy spectrum has a vertical tangent at $\varepsilon \to \varepsilon_m$ for $m_\nu \neq 0$ and a horizontal tangent for $m_\nu = 0$. (It is this difference which enables an estimate to be made of m_ν, for example in the low energy β-decay of ^{3}H discussed in section 6.5.4).

6.10 The atomic masses of ^{7}Be and ^{7}Li are 7.0169 u and 7.0160 u respectively. Show that ^{7}Be can only decay by electron capture. What is the lowest value for the mass of ^{7}Be which would allow β^+-decay?

6.11 Using the same approach as in section 6.5.2 show that the selection rules for a first forbidden β-decay $(L = 1)$ are as follows

| Fermi transition (V) | $\Delta J = 0, \pm 1$ | $\Delta \pi =$ 'yes' |
| Gamow–Teller transition (A) | $\Delta J = 0, \pm 1, \pm 2$ | $\Delta \pi =$ 'yes' |

Which transitions satisfying these selection rules are nevertheless not permitted by angular momentum conservation?

7 Elementary particles and their interactions

7.1 Refer to Table 7.1.

(a) From the value given for the width for W boson decay estimate an approximate limit for the corresponding mean life.

(b) Investigate the change in isospin, ΔI, implied by the weak hadronic decays of the Λ, Σ, Ξ and K particles. What is the common feature?

(c) By considering the charge (Q – in units of e), the 3-component of isospin (I_3), baryon number (B) and strangeness (S) for the Λ^0, K^- and Ω^-, determine the parameters a, b and c in the expression

$$Q = aI_3 + bB + cS$$

and check that the result agrees with eq. (7.3).

7.2 Find the allowed values of the total isospin for the final states in the following strong interaction processes:

(a)
$$d + p \rightarrow n + p + p$$

(b)
$$p + \pi^- \rightarrow n + \pi^0$$

(c)
$$p + \pi^+ \rightarrow p + \pi^+$$

(d)
$$p + \pi^+ \rightarrow n + \pi^+ + \pi^+$$

7.3 Which of the following processes are allowed strong interaction processes? For those that are not allowed state which conservation law is violated.

(a)
$$p + \pi^- \rightarrow \pi^0 + n$$

(b)
$$K^0 + p \rightarrow \Sigma^+ + \pi^0$$

(c)
$$\pi^0 + n \rightarrow \pi^+ + \pi^-$$

(d)
$$K^- + p \rightarrow \Lambda^0 + n$$

(e)
$$\pi^- + p \rightarrow \Lambda^0 + K^0$$

7.4 A new particle X is discovered which decays weakly as follows

$$X \rightarrow \pi^0 + \mu^+$$

Determine the following properties of X:

(a) lepton number;
(b) baryon number;
(c) charge;
(d) whether it is a boson or a fermion;
(e) a lower limit on its mass in units of MeV/c^2.

What would be the ultimate decay products in the above process?

7.5 In the following processes indicate, with an explanation, whether they proceed by the strong, electromagnetic or weak interaction or whether they cannot occur.

(a)
$$\pi^+ \to \mu^+ + \nu_\mu$$

(b)
$$p \to n + e^+ + \nu_e$$

(c)
$$\Lambda^0 \to p + \pi^-$$

(d)
$$\pi^+ \to \gamma + \gamma$$

(e)
$$p + \pi^- \to \Lambda^0 + \pi^0$$

(f)
$$p + \pi^- \to K^+ + \Sigma^-$$

(g)
$$n + \pi^+ \to \Sigma^- + p$$

7.6 Explain why the reaction

$$\pi^- + {}^2H \to 2n + \pi^0$$

is not observed for pions captured at rest.

8 The strong interaction

8.1 Apply the arguments used in section 8.2 to show that the magnetic moment, μ_Λ, of the Λ^0 is simply the magnetic moment, μ_s, of the s quark. Hence, by writing

$$\mu_s = g \frac{Q_s}{2m_s} s_s$$

as in eq. (8.3), where g is the quark g factor, Q_s is the s quark charge and m_s is its constituent mass, show that

$$\mu_\Lambda = - \frac{m}{3m_s} \mu_p$$

where μ_p is the proton magnetic moment and m is the u, d quark constituent mass.

Experimentally it is found that $\mu_\Lambda \simeq - 0.61 \mu_N$. Determine the ratio m_s/m and compare it with the value which can be derived from Table 8.3.

8.2 The following charmed baryons have been identified:

$$\Sigma_c^{++}, \Sigma_c^{+}, \Sigma_c^{0} \text{ with } S = 0 \text{ and masses} \simeq 2450 \,\mathrm{MeV}/c^2$$
$$\Xi_c^{+} \text{ with } S = - 1 \text{ and mass} \simeq 2460 \,\mathrm{MeV}/c^2$$

Speculate on their likely quark content.

8.3 The charmed mesons $D^{+,0}$ have $J^\pi = 0^-$ and their excited states $D^{*+,0} (2010)$ have $J^\pi = 1^-$. Identify their quark content and give the values of total quark spin and orbital angular momentum.

9 The weak interaction and unification

9.1 For the decay $\Lambda^0 \to \pi^- + p$ determine the possible values for the orbital angular momentum in the final state. Which of these values is allowed if parity is conserved?

The observed angular distribution of the pion in the centre-of-mass frame has the form $I(\theta) \propto 1 + \kappa \cos \theta$ where θ is the angle between the direction of motion of the pion and the spin of the Λ^0 particle. Verify that such an angular distribution is incompatible with the conservation of parity.

9.2 In section 9.3.1 the observation of the process

$$\tilde{\nu}_\mu + e^- \to \tilde{\nu}_\mu + e^-$$

was taken to indicate the presence of a neutral current interaction. Why does the process

$$\tilde{\nu}_e + e^- \to \tilde{\nu}_e + e^-$$

not similarly indicate the presence of such an interaction?

9.3 The α-decay of an excited 2^- state in ^{16}O to the ground state (0^+) of ^{12}C has been observed with a width $\Gamma_\alpha \simeq 1 \times 10^{-10} \,\mathrm{eV}$. Explain why this indicates the presence of a parity-violating potential.

9.4 Experiments have been performed to measure the average values of the following quantities for polarized neutrons and their β-decays: (a) $p_c \cdot J$; (b) $p_v \cdot J$; (c) $(p_v \times p_c) \cdot J$.

The p vectors signify the final momenta of the electron or neutrino and J is the neutron spin operator. What forms do each of these operators take after being subject once to the parity operator P or once to the time reversal operator T? For each operator what would be the significance with respect to these symmetries of a non-zero result for the measurement?

Answers to problems

1 The nuclear atom

1.1 For Ag,
$$d_0 = 2.71 \times 10^{-14} \, \text{m at 5 MeV}$$
$$= 1.93 \times 10^{-14} \, \text{m at 7 Mev}$$

For Au,
$$d_0 = 4.55 \times 10^{-14} \, \text{m at 5 MeV}$$
$$= 3.25 \times 10^{-14} \, \text{m at 7 MeV}$$

1.2 (i) Electron kinetic energy $\simeq 50 \, \text{MeV}$; (ii) proton kinetic energy $\simeq 2 \, \text{MeV}$.

1.3

A	Z	Spin (ep)	Spin (np)
even	even	integer	integer
even	odd	half-integer	integer
odd	even	integer	half-integer
odd	odd	half-integer	half-integer

2 General properties of the nucleus

2.2 $B = 160.8 \, \text{MeV}$.

2.3 (i) $E_n = 15.65 \, \text{MeV}$; (ii) $E_p = 12.11 \, \text{MeV}$.

3 The internucleon potential

3.1 $V \simeq 10 \, \text{MeV}$.

3.2 $A = -30\,\text{MeV}$; $B\hbar^2 = -20\,\text{MeV}$.

3.3 $R \simeq 4.2\,\text{fm}$.

3.4 (i) 0.36 fm; (ii) 0.26 fm; (iii) 0.37 fm.

4 Models of nuclear structure

4.3 ^{15}N: $(1p_{1/2})^1$
^{21}Ne: $(3d_{5/2})^3$ coupled to $J^P = \frac{3}{2}^+$
^{39}K: $(1d_{3/2})^5$
^{51}V: $(1f_{7/2})^3$
^{73}Ge: $(1g_{9/2})^1$
^{207}Pb: $(3p_{1/2})^1$ – odd nucleon not in $1i_{13/2}$ state (pairing)

4.4 ^{113}In: $(1g_{9/2})^{-1}$
^{121}Sb: $(2d_{5/2})^1$ – spin and magnetic moment indicate
odd nucleon in $2d_{5/2}$ state rather than $1g_{7/2}$
^{123}Sb: $(1g_{7/2})^1$
^{123}Te: $(3s_{1/2})^1$ – spin and magnetic moment indicate
odd nucleon in $3s_{1/2}$ state rather than $1h_{11/2}$ (pairing)

4.5 $\lambda = -2\,\text{MeV}$.

4.7 $E_{6^+} \simeq 0.63\,\text{MeV}$ (cf. experiment gives 0.59 MeV); $I/I_{\text{rigid}} \simeq 0.45$.

5 Nuclear reactions

5.1 Proton energy = 3.28 MeV.

5.2 Energy of ^{16}O ions = 10.4 MeV. Orbital angular momentum $\simeq 19\hbar$.

5.3 Foil thickness = $4.96 \times 10^{-5}\,\text{m}$.

5.4

$$^{35}_{17}\text{Cl} + \text{p} \rightarrow {}^{36}_{18}\text{Ar}^* \rightarrow {}^{32}_{16}\text{Si} + {}^4_2\text{He}$$
$$^{10}_{5}\text{B} + \text{n} \rightarrow {}^{11}_{5}\text{B}^* \rightarrow {}^7_3\text{Li} + {}^4_2\text{He}$$
$$^{23}_{11}\text{Na} + \text{p} \rightarrow {}^{24}_{12}\text{Mg}^* \rightarrow {}^{20}_{10}\text{Ne} + {}^4_2\text{He}$$
$$^{23}_{11}\text{Na} + {}^2_1\text{H} \rightarrow {}^{25}_{12}\text{Mg}^* \rightarrow {}^{24}_{11}\text{Na} + \text{p}$$

5.5 Neutron release rate = 4.6×10^{18} neutrons s^{-1}. Longest period of continuous running $\simeq 42$ days.

6 Alpha, beta and gamma decay

6.1 Activity = 7.13×10^6 decays s^{-1}.

6.2 16.6% decays in 1 day and 83.7% in 10 days.

6.4 Kinetic energy $= 5.31\,\text{MeV}$.

6.6

$$\tfrac{7}{2}^+ \to \tfrac{5}{2}^+ \ (\text{M1}(+\text{E2})), \ \tfrac{7}{2}^+ \to \tfrac{11}{2}^- \ (\text{M2}),$$
$$\tfrac{5}{2}^+ \to \tfrac{1}{2}^+ \ (\text{E2}), \ \tfrac{11}{2}^- \to \tfrac{1}{2}^+ \ (\text{E5})$$

6.7 (a) E2; (b) $15.8\,\text{eV}$; (c) $\Gamma \simeq 0.1\,\text{eV}$ – no resonant absorption; (d) energy shift due to recoil negligible and resonant absorption possible.

6.9 For $m_v = 0$, $P(\varepsilon) \propto x^2$ and slope $\propto x \to 0$ for $x \to 0$. For $m_v \neq 0$, $P(\varepsilon) \propto x^{1/2}$ and slope $\propto x^{-1/2} \to \infty$ for $x \to 0$.

6.10 Lowest mass for $^7\text{Be} = 7.0171\,\text{u}$.

6.11 Fermi transition, $0 \to 0$ not allowed.

7 Elementary particles and their interactions

7.1 (a) $\tau \geqslant 10^{-25}\,\text{s}$; (b) for all decays $\Delta I = $ half-integer.

7.2 (a) $I = 1/2$; (b) $I = 1/2, 3/2$; (c) $I = 3/2$; (d) $I = 3/2$.

7.3 (a) Allowed; (b) forbidden $(\Delta S \neq 0)$; (c) forbidden $(\Delta B \neq 0)$; (d) forbidden $(\Delta B \neq 0)$; (e) allowed.

7.4 (a) $L_\mu = -1$, $L_e = 0$; (b) $B = 0$; (c) $Q = e$; (d) fermion; (e) $M_X \geqslant 240.6\,\text{MeV}$.

7.5 (a) Weak; (b) forbidden $(m_p < m_n)$; (c) weak $(|\Delta S| = 1)$; (d) forbidden $(\Delta Q \neq 0)$; (e) weak $(|\Delta S| = 1)$; (f) strong $(\Delta S = 0)$; (g) forbidden $(\Delta Q \neq 0)$.

7.6 Parity not conserved.

8 The strong interaction

8.1 $m_s/m = 1.52$. Table 8.3 gives 1.42.

8.2 Σ_c^{++}(uuc), Σ_c^+(udc), Σ_c^0(ddc), Ξ_c^+(usc).

8.3 D^+ (c$\tilde{d}$), D^0 (c$\tilde{u}$) have $S = 0$, $L = 0$; D^{*+} (c$\tilde{d}$), D^{*0}(c$\tilde{u}$) have $S = 1$, $L = 0$.

9 The weak interaction and unification

9.1 $l = 0$ or (a) $l = 1$ conserves parity.

$I(\theta) \propto 1 + \kappa\, \boldsymbol{J}\cdot\boldsymbol{p}/Jp$ where $\boldsymbol{J}$ is the Λ^0 spin vector and $\boldsymbol{p}$ is the pion momentum. $\boldsymbol{J}\cdot\boldsymbol{p}$ is a pseudoscalar implying parity non-conservation (cf. parity non-conservation in β-decay, section 6.5.3).

9.2 The process $\tilde{\nu}_e + e^- \rightarrow \tilde{\nu}_e + e^-$ can be propagated by the interaction between charged leptonic currents.

9.3 Emitted α would have $l = 2$ (even parity) whereas a parity change is needed for a $2^- \rightarrow 0^+$ transition.

9.4 (a) P odd and T even; (b) P odd and T even; (c) P even and T odd. A non-zero measurement implies P violation in (a) and (b) and T violation in (c).

Index